ELECTRICAL ENGINEERING
QUICK REFERENCE CARDS

Kenneth A. Nelson, P.E.

PROFESSIONAL PUBLICATIONS, INC.
Belmont, CA 94002

In the ENGINEERING REFERENCE MANUAL SERIES

Engineer-In-Training Reference Manual
    Engineering Fundamentals Quick Reference Cards
    Engineer-In-Training Sample Examinations
    Mini-Exams for the E-I-T Exam
    1001 Solved Engineering Fundamentals Problems
    E-I-T Review: A Study Guide
Civil Engineering Reference Manual
    Civil Engineering Quick Reference Cards
    Civil Engineering Sample Examination
    Civil Engineering Review Course on Cassettes
    Seismic Design of Building Structures
    Seismic Design Fast
    Timber Design for the Civil P.E. Exam
Structural Engineering Practice Problem Manual
Mechanical Engineering Reference Manual
    Mechanical Engineering Quick Reference Cards
    Mechanical Engineering Sample Examination
    101 Solved Mechanical Engineering Problems
    Mechanical Engineering Review Course on Cassettes
    Consolidated Gas Dynamics Tables
Electrical Engineering Reference Manual
    Electrical Engineering Quick Reference Cards
    Electrical Engineering Sample Examination
Chemical Engineering Reference Manual
    Chemical Engineering Practice Exam Set
Land Surveyor Reference Manual
Petroleum Engineering Practice Problem Manual
Expanded Interest Tables
Engineering Law, Design Liability, and Professional Ethics
Engineering Unit Conversions

In the ENGINEERING CAREER ADVANCEMENT SERIES

How to Become a Professional Engineer
The Expert Witness Handbook—A Guide for Engineers
Getting Started as a Consulting Engineer
Intellectual Property Protection—A Guide for Engineers
E-I-T/P.E. Course Coordinator's Handbook
Becoming a Professional Engineer

ELECTRICAL ENGINEERING QUICK REFERENCE CARDS

Copyright © 1991 by Professional Publications, Inc. All rights are reserved. No part of this publication may be reproduced, stored in a retrieval system, or transmitted, in any form or by any means, electronic, mechanical, photocopying, recording, or otherwise, without the prior written permission of the publisher.

Printed in the United States of America

ISBN: 0-912045-21-3

Professional Publications, Inc.
1250 Fifth Avenue, Belmont, CA 94002
(415) 593-9119

Current printing of this edition (last number): 6 5 4 3 2 1

TABLE OF CONTENTS

MATHEMATICS 1

LINEAR CIRCUIT ANALYSIS 6

WAVEFORMS, POWER, AND MEASUREMENTS 11

TIME AND FREQUENCY RESPONSE 13

POWER SYSTEMS 18

TRANSMISSION LINES 21

ROTATING MACHINES 24

FUNDAMENTAL SEMICONDUCTOR CIRCUITS 28

AMPLIFIER APPLICATIONS 37

WAVESHAPING, LOGIC, AND DATA CONVERSION 44

DIGITAL LOGIC 51

CONTROL SYSTEMS 57

ECONOMICS 63

MATHEMATICS

LINEAR EQUATIONS

Given two coordinate points (x_1, y_1) and (x_2, y_2), x and y are satisfied by the following equation:

$$m = \frac{y - y_1}{x - x_1} = \frac{y_2 - y_1}{x_2 - x_1}$$
$$= \text{slope of line}$$

An alternate form is $y = mx + b$, where $y = b$ at $x = 0$.

LINEAR SIMULTANEOUS EQUATIONS

$$a_{11}x_1 + a_{12}x_2 + a_{13}x_3 + \cdots + a_{1n}x_n = y_1$$
$$a_{21}x_1 + a_{22}x_2 + a_{23}x_3 + \cdots + a_{2n}x_n = y_2$$
$$\cdots + \cdots + \cdots + \cdots + \cdots = \cdots$$
$$a_{n1}x_n + a_{n2}x_2 + a_{n3}x_3 + \cdots + a_{nn}x_n = y_n$$

Use Cramer's rule to solve for y_n.

$$M = \text{coefficient matrix}$$
$$= \begin{vmatrix} a_{11} & a_{12} & \ldots & a_{1n} \\ a_{21} & a_{22} & \ldots & a_{2n} \\ \vdots & & \ddots & \vdots \\ a_{n1} & a_{n2} & \ldots & a_{nn} \end{vmatrix}$$

$$y_n = \frac{N}{D}$$

1. Make a numerator determinant, N, from the coefficient matrix, M, by substituting elements from the column matrix, Y, into the nth column of the coefficient matrix.

2. Make a denominator determinant, D, from the coefficient matrix.

3. Find the value of the determinants by the sum of the products of any row or column elements times each element cofactor, where i = row and j = column.

$$\text{cofactor of elements } a_{ij} = A_{ij} = (-1)^{1+j}\text{minor}_{ij}$$

$$\text{determinant value} = \sum_{i=1}^{i=n} a_{ij}A_{ij} \quad j = 1 \text{ or } 2 \text{ or } \ldots n$$

A minor is the determinant formed by striking out the ith row and jth column containing the element.

4. Continue reducing minors until a second-order minor is found. Then use the formula

$$\begin{vmatrix} A & B \\ C & D \end{vmatrix} = AD - BC$$

QUADRATIC EQUATIONS

$$ax^2 + bx + c = y$$

a, b, and c are constants. For $y = 0$, the solution to x is

$$x_{1,2} = \frac{-b \pm \sqrt{b^2 - 4ac}}{2a}$$

POLYNOMIALS

A polynomial of form

$$x^n + a_1 x^{n-1} + a_2 x^{n-2} + \cdots + a_{n-1} x^n + a_n = 0$$

has the factored form

$$(x + p_1)(x + p_2)(x + p_3) \cdots (x + p_n) = 0$$

p_n are the roots and can be real or complex.

Use a formula to solve for roots when $n = 1$ or 2. Use factoring for simple forms.

$$(a + b)(c + d) = ac + ad + bc + bd$$
$$(a + b)^2 = a^2 + 2ab + b^2$$
$$(a + b)^3 = a^3 + 3a^2b + 3ab^2 + b^3$$

Plot the equation and interpolate the intersection(s) of the curve with the x-axis where $y = 0$.

Use Steinman's method for unfactorable polynomials. Let the first guess be x. The first approximation is

$$x_1 = \frac{-a_1 - \dfrac{2a_2}{x} - \dfrac{3a_3}{x^2} - \dfrac{4a_4}{x^3} - \cdots}{1 - \dfrac{a_2}{x^2} - \dfrac{2a_3}{x^3} - \dfrac{3a_4}{x^4} \cdots}$$

Substitute x_1 for x to find the second approximation, x_2. Continue the iteration until the selected significant digit of x no longer changes.

PARTIAL FRACTION EXPANSION

The fraction $h(x) = f(x)/g(x)$ can be broken down into the sum of partial fractions using the following rules:

1. If the degree of $f(x)$ is greater than the degree of $g(x)$, divide $g(x)$ into $f(x)$ to get a quotient plus a remainder whose numerator is of a lesser degree than its denominator.

$$h(x) = \frac{x^3 + 2x^2 + x + 6}{x^2 + 2x - 3} = \frac{f(x)}{g(x)}$$
$$= x + \frac{4x + 6}{x^2 + 2x - 3}$$
$$= x + \frac{d(x)}{g(x)}$$

2. Factor $g(x)$.
$$x^2 + 2x - 3 = (x+3)(x-1)$$

3. Set $\dfrac{d(x)}{g(x)} = \dfrac{A}{x+3} + \dfrac{B}{x-1}$.

4. Solve for arbitrary constants A and B by multiplying by $g(x)$ and combining like terms.
$$d(x) = \left(\dfrac{A}{x+3} + \dfrac{B}{x-1}\right)[(x+3)(x-1)]$$
$$4x + 6 = A(x-1) + B(x+3)$$
$$Ax + Bx = 4x$$
$$-A + 3B = 6$$
$$B = 2.5 \text{ and } A = 1.5$$

5. Rewrite the fraction using the constants.
$$h(x) = x + \dfrac{1.5}{x+3} + \dfrac{2.5}{x-1}$$

6. For factors to a power, each power factor must be a partial fraction.
$$\dfrac{f(x)}{(x+k)^n} = \dfrac{A}{x+k} + \dfrac{B}{(x+k)^1} + \cdots$$

7. For a quadratic equation, use x in the numerator as follows:
$$\dfrac{f(x)}{x^2 + px + q} = \dfrac{Bx + C}{x^2 + px + q}$$

8. For combinations of factors, combine all the above rules with each factor and power factor being a partial fraction.

TRIGONOMETRY

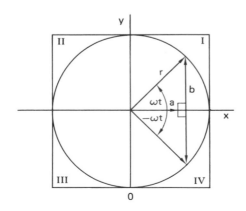

The sides of a right triangle can be defined by the following:
$$x^2 + y^2 = r^2$$
$$\sin^2 \omega t + \cos^2 \omega t = 1$$
$$\sin \omega t = \dfrac{b}{r}$$
$$\csc \omega t = \dfrac{r}{b} = \dfrac{1}{\sin \omega t}$$
$$\cos \omega t = \dfrac{a}{r}$$
$$\sec \omega t = \dfrac{r}{a} = \dfrac{1}{\cos \omega t}$$
$$\tan \omega t = \dfrac{b}{a} = \dfrac{\sin \omega t}{\cos \omega t}$$
$$\cotan \omega t = \dfrac{a}{b} = \dfrac{1}{\tan \omega t}$$

The sign of the trigonometric function can be found by an inspection of the angle and the quadrant.
$$\sin(-\omega t) = -\sin \omega t$$
$$\cos(-\omega t) = \cos \omega t$$
$$\tan(-\omega t) = -\tan \omega t$$

The angle is positive in the counterclockwise direction and negative in the clockwise direction.

The sum of the internal angles of a right triangle is 180°. To find the sine or cosine of multiple angles use the trigonometric identities for multiple angles.
$$\pi \text{ radians} = 180°$$

Important Trigonometric Identities

$$\sin(x \pm y) = \sin x \cos y \pm \cos x \sin y$$
$$\cos(x \pm y) = \cos x \cos y \mp \sin x \sin y$$
$$\sin 2x = 2 \sin x \cos x$$
$$\sin^2 \omega t + \cos^2 \omega t = 1$$

Law of Cosines

For any triangle,
$$a^2 = b^2 + c^2 - 2bc(\cos A)$$

A = included angle between sides b and c.

EXPONENTS

If n and m are real, fractional, or complex numbers,
$$V^n V^m = V^{n+m}$$
$$\dfrac{V^n}{V^m} = V^{n-m}$$

MATHEMATICS

LOGARITHMS

A logarithm is an exponent.

$$10^n = Y, \text{ then } \log_{10} Y = n$$

$$e^n = Y, \text{ then } \ln Y = n \quad (e = 2.71828\ldots)$$

10 and e are the bases of the logarithm.

$$\log_a x = \frac{\ln_e x}{\ln_e a} \quad (a = \text{any base})$$

$$\log_a xy = \log_a x + \log_a y$$

$$\log_a \left(\frac{x}{y}\right) = \log_a x - \log_a y$$

$$\log_a(x^n) = n \log_a x$$

COMPLEX NUMBERS

$$j = \sqrt{-1}$$
$$j^2 = -1$$
$$j^3 = -\sqrt{-1}$$
$$j^4 = 1$$

Rectangular Form

$$V_1 = a_1 + jb_1 \quad V_2 = a_2 + jb_2$$
$$E_1 = \text{constant} \quad E_2 = \text{constant}$$

Addition and Subtraction

$$V_1 \pm V_2 = (a_1 + jb_1) \pm (a_2 + jb_2)$$
$$= (a_1 \pm a_2) + j(b_1 \pm b_2)$$

Multiplication

$$V_1 V_2 = (a_1 + jb_1)(a_2 + jb_2)$$
$$= (a_1 a_2 - b_1 b_2) + j(a_1 b_2 + b_1 a_2)$$

Division

Use the polar form.

$$\frac{V_1}{V_2} = \left(\frac{a_1 + jb_1}{a_2 + jb_2}\right)\left(\frac{a_2 - jb_2}{a_2 - jb_2}\right)$$

$$\frac{V_1}{V_2} = \left[\frac{(a_1 a_2 + b_1 b_2) + j(b_1 a_2 - a_1 b_2)}{(a_2)^2 + (b_2)^2}\right]$$

Exponential Form

$$e^{\pm j\omega t} = \cos \omega t \pm j \sin \omega t$$
$$E_1 e^{j\omega t} = E_1(\cos \omega t + j \sin \omega t)$$

Addition or subtraction:

Convert to rectangular form, perform operation, then convert back to exponential form.

Multiplication:

$$(E_1 e^{j\omega t_1})(E_2 e^{j\omega t_2}) = E_1 E_2 e^{j(\omega t_1 + \omega t_2)}$$
$$= E_1 E_2 [\cos(\omega t_1 + \omega t_2) + j \sin(\omega t_1 + \omega t_2)]$$

Division:

$$\frac{E_1 e^{j\omega t_1}}{E_2 e^{j\omega t_2}} = \frac{E_1}{E_2} e^{j(\omega t_1 - \omega t_2)}$$
$$= E[\cos(\omega t_1 - \omega t_2) + j \sin(\omega t_1 - \omega t_2)]$$

Polar Form

The polar form is a shorthand way of writing the exponential form.

$$E e^{j\omega t} = E \underline{/\omega t}$$
$$E = \text{magnitude}$$
$$\omega t = \text{angle in degrees}$$

Note: Be careful of radians and degrees. ωt in $e^{j\omega t}$ is in radians.

Addition and subtraction:

Convert to rectangular form, perform operation, then convert back to polar form.

Multiplication:

$$E_1 \underline{/\omega t_1} \ E_2 \underline{/\omega t_2} = E_1 E_2 \underline{/\omega t_1 + \omega t_2}$$

Division:

$$\frac{E_1 \underline{/\omega t_1}}{E_2 \underline{/\omega t_2}} = \frac{E_1}{E_2} \underline{/\omega t_1 - \omega t_2}$$

COMPLEX NUMBER EQUIVALENCE

Two phasors are equivalent when their real and imaginary parts are equal.

$$V_1 = a_1 + jb_1$$
$$V_2 = a_2 + jb_2$$
$$V_1 = V_2$$

The following forms are also equivalent:

$$V = a + jb = \sqrt{a^2 + b^2} \underline{/\arctan(b/a)}$$
$$= \sqrt{a^2 + b^2} \exp[(\pi/180) \arctan(b/a)]$$

$\exp(n) = e^n$, and $\arctan(b/a)$ is in degrees.

COMPLEX CONJUGATE

The complex conjugate of $V = a + jb$ is $V^* = a - jb$. $VV^* = $ real number $= a^2 + b^2$.

ROOTS AND POWERS

Evaluate roots and powers using the polar or exponential form and apply the rules of exponents.

$$V^n = |V|^n \underline{/n\omega t}$$
$$n = \text{real, whole, or fraction}$$

DIFFERENTIALS

$$du^a = au^{a-1}du$$
$$de^u = e^u du$$
$$d(\text{constant}) = 0$$
$$d(\sin \omega t) = \omega \cos \omega t$$
$$d(\cos \omega t) = -\omega \sin \omega t$$
$$d(uv) = udv + vdu$$
$$d\ln u = \frac{du}{u}$$

INTEGRALS

$$\int u^a du = \frac{u^{a+1}}{a+1} + C \quad (a \neq -1)$$
$$\int e^u du = e^u + C$$
$$\int \sin \omega t \, d\omega t = -\frac{1}{\omega} \cos \omega t + C$$
$$\int \cos \omega t \, d\omega t = \frac{1}{\omega} \sin \omega t + C$$
$$\int u dv = uv - \int v du$$
$$\int e^{at} \cos \omega t \, d\omega t = \left(\frac{e^{at}}{a^2 + \omega^2}\right)(a \cos \omega t + b \sin \omega t) + C$$
$$\int e^{at} \sin \omega t \, d\omega t = \left(\frac{e^{at}}{a^2 + \omega^2}\right)(a \sin \omega t - b \cos \omega t) + C$$

If $f''(u) = 0$, there is an inflection at u.
If $f''(u)$ is positive, there is a minimum at u.
If $f''(u)$ is negative, there is a maximum at u.

There can be multiple combinations of maximums, minimums, or inflections.

LINEAR FIRST-ORDER DIFFERENTIAL EQUATIONS

$$bf'(x) + cx = d$$

The solution is

$$x = \frac{d}{c} + Ae^{\frac{-ct}{b}}$$

LINEAR SECOND-ORDER DIFFERENTIAL EQUATIONS

$$af''(x) + bf'(x) + cx = d$$

Case 1: Overdamping

When $b^2 - 4ac$ is positive,

$$x = \frac{d}{c} + Ae^{-g_1 t} + Be^{-g_2 t}$$

where $g_1 = \frac{b + \sqrt{b^2 - 4ac}}{2a}$

$$g_2 = \frac{b - \sqrt{b^2 - 4ac}}{2a}$$

A and B are constants whose values depend on the initial conditions when $t = 0$.

Case 2: Critically Damped

When $b^2 - 4ac$ is zero,

$$x = \frac{d}{c} + Ae^{-\frac{bt}{2a}} + Bte^{-\frac{bt}{2a}}$$

Case 3: Underdamped

When $b^2 - 4ac$ is negative,

$$x = \frac{d}{c} + Ae^{-\alpha t} \cos \beta t + Be^{-\alpha t} \sin \beta t$$
$$\alpha = \frac{b}{2a}$$
$$\beta = \frac{\sqrt{4ac - b^2}}{2a}$$

FOURIER SERIES

A repeating waveform of period T can be expressed by the following infinite series:

$$f(t) = A_0 + \sum_{n=1}^{\infty} a_n \cos\left(\frac{2\pi nt}{T}\right) + \sum_{n=1}^{\infty} b_n \sin\left(\frac{2\pi nt}{T}\right)$$

$$A_0 = \frac{1}{T} \int_{t_1}^{t_1 + T} f(t) dt$$

A_0 is also the average value of the wave.

$$a_n = \frac{2}{T}\int_{t_1}^{t_1+T} f(t)\cos\left(\frac{2\pi nt}{T}\right)dt$$

$$b_n = \frac{2}{T}\int_{t_1}^{t_1+T} f(t)\sin\left(\frac{2\pi nt}{T}\right)dt$$

There are no sine terms for even functions. An even function is when $f(t) = f(-t)$. The cosine is an even function.

There are no cosine terms for odd functions. An odd function is when $f(t) = -f(-t)$. The sine is an odd function.

FOURIER TRANSFORM PAIR

Transient analysis involves non-periodic waveforms that are not subject to Fourier series. By applying a limit to a periodic wave as the period approaches infinity, the Fourier transform pair is obtained. This transforms any waveform from a function of frequency to a function of time.

$$f(t) = \frac{1}{2\pi}\int_{-\infty}^{\infty} g(\omega)e^{j\omega t}d\omega$$

$$g(\omega) = \int_{-\infty}^{\infty} f(t)e^{-j\omega t}dt$$

LAPLACE TRANSFORM INTEGRAL

$g(\omega)$ cannot be evaluated for some waveforms, so a converging factor, e^{-at}, is applied to the $g(\omega)$ integral and a limit applied as a goes to 0. A new variable, $s = a + j\omega$, is defined. $g(\omega)$ now becomes $F(s)$ and $f(t)e^{-j\omega t}$ becomes $f(t)e^{-st}$.

$$F(s) = \int_0^{\infty} f(t)e^{-st}dt = \mathcal{L}f(t)$$

To change back from the Laplace (s domain) to the time domain, apply the following integral.

$$f(t) = \frac{1}{2\pi j}\int_{a-j\infty}^{a+j\infty} F(s)e^{st}ds$$

LAPLACE THEOREMS

1. $\mathcal{L}[f_1(t) + f_2(t)] = \mathcal{L}[f_1(t)] + \mathcal{L}[f_2(t)]$
2. $\mathcal{L}[af(t)] = a\mathcal{L}[f(t)]$
3. $\mathcal{L}(e^{-at})f(t) = F(s+a) = \mathcal{L}[f(t)]$ at $s = s+a$
4. If $\mathcal{L}[f(t)] = F(s)$,
   $\mathcal{L}[f(t-t_0)]U(t-t_0) = e^{-t_0 s}F(s)$
5. The Laplace of a periodic function is K times the Laplace of the first cycle. $K = 1/1 - e^{-Ts}$ where T = period of waveform.
6. $\mathcal{L}[f'(t)] = sF(s) - f(0)$
   $f(0)$ = value of $f(t)$ at $t = 0$
7. $\mathcal{L}[f''(t)] = s^2 F(s) - sf(0) - f'(0)$
   $f(0)$ = value of $f(t)$ at $t = 0$
   $f'(0)$ = value of $f'(t)$ at $t = 0$
8. $\mathcal{L}[\int_0^t f(t)dt] = \frac{F(s)}{s}$

   $\mathcal{L}\left[\frac{f(t)}{t}\right] = \int_s^{\infty} F(s)ds$

9. Initial Value Theorem

   $f(0) = \text{limit } f(t)$ as $t \to 0^+$
   $= \text{limit } sF(s)$ as $s \to \infty$

10. Final Value Theorem

    $f(\infty) = \text{limit } f(t)$ as $t \to \infty$
    $= \text{limit } sF(s)$ as $s \to 0^+$

LAPLACE TRANSFORMS

impulse function, $\delta(t)$	1

Note: The impulse function has zero width, infinite height, and an area of 1.

unit step function, $u(t)$	$\frac{1}{s}$
constant a	$\frac{a}{s}$
$f(t)$	$F(s)$
e^{-at}	$\frac{1}{s+a}$
te^{-at}	$\frac{1}{(s+a)^2}$
$\sin bt$	$\frac{b}{s^2+b^2}$
$\cos bt$	$\frac{s}{s^2+b^2}$
$e^{-at}\sin bt$	$\frac{b}{(s+a)^2+b^2}$
$e^{-at}\cos bt$	$\frac{s+a}{(s+a)^2+b^2}$
$\frac{t^{n-1}}{n-1}$	$\frac{1}{s^n}$
t^n	$\frac{n!}{s^{n+1}}$
$\sinh bt$	$\frac{b}{s^2-b^2}$
$\cosh bt$	$\frac{s}{s^2-b^2}$

Note: $n!$ means factorial of n.

$$0! = 1 \qquad 5! = (5)(4)(3)(2)(1)$$

LINEAR CIRCUIT ANALYSIS

CIRCUIT ELEMENTS

Ideal Voltage Source

DC: v = constant
AC: $v = E_{max} e^{j\omega t} = E_{max}(\cos \omega t + j \sin \omega t)$
Z (internal) = 0

Ideal Current Source

DC: i = constant
AC: $i = I_{max} e^{j\omega t} = I_{max}(\cos \omega t + j \sin \omega t)$
Z (internal) = ∞

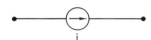

Dependent sources are the same as ideal sources except the voltage or current values are functions of other circuit variables.

Resistance

Ohm's law is $R = v/i$ in Ω.
Conductance is $G = 1/R$ in S.

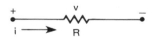

DC power converted to heat in W is

$$P = \frac{v^2}{R} = i^2 R = iv$$

AC power converted to heat in W is

$$P = |v| |i| \cos \theta$$
θ = power factor angle

Resistance is a function of temperature given by

$$\rho_T = \rho_{20}[1 + \alpha_{20}(T - 20)]$$

$$R = \frac{\rho_T l}{A} \quad l = \text{length}, \; A = \text{area}, \; T = °C$$

ρ_{20} = resistivity at 20°C
α_{20} = temperature coefficient of resistivity at 20°C

	copper	nichrome
ρ_{20}	1.8×10^{-8} $\Omega \cdot$m	1.08×10^{-6} $\Omega \cdot$m
α_2	$3.9 \times 10^{-3}/°C$	$17 \times 10^{-5}/°C$

4 c.m. = π square mils

Capacitance

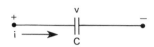

$$C = \frac{q}{v} \text{ in F}$$

q = charge in C
V = voltage in V
power converted to heat = 0
VA reactive power: $Q = |v||i| \sin \theta = |v||i|$
θ = power factor angle = 90°
stored electrical energy: $W = \frac{1}{2} C v^2$ J

voltage: $v = \frac{q}{C} = \frac{1}{C} \int i \, dt + V_0$

$v = i(-jX_C)$

current: $i = C \frac{dv}{dt}$

$\omega = 2\pi f$ in rad/s

f = frequency in Hz

reactance: $X_C = \dfrac{1}{\omega C}$ in Ω

susceptance: $\dfrac{1}{X_C} = B$ in S

$C = \epsilon_o \epsilon_r \dfrac{A}{d}$

ϵ_o = permittivity of free space
$= \dfrac{1}{36\pi} \times 10^{-9}$ F/m

ϵ_r = relative permittivity of material
(varies from 1 to 10 for solids)

Inductance

$L = \dfrac{N\phi}{i}$ in H

$N\phi$ = flux linkages in Wb

power converted to heat = 0

LINEAR CIRCUIT ANALYSIS

VA reactive power: $Q = |v||i|\sin\theta = |v||i|$

θ = power factor angle = 90°

stored electrical energy: $W = \dfrac{Li^2}{2}$ in J

voltage: $v = L\dfrac{di}{dt} = i(jX_L)$

current: $i = \dfrac{1}{L}\int v\, dt + I_0$

$\omega = 2\pi f$ in rad/s

f = frequency in Hz

reactance: $X_L = \omega L$ in Ω

susceptance: $B = \dfrac{1}{X_L}$ in S

A node is a connection point of any number of elements. Voltage drop has a sign of + to − in the direction of current flow through a resistance, capacitance, or inductance. Voltage rise has a sign of − to + in the direction of current flow through a source. Conventional current flow is from + to −.

impedance: $Z = R + j(X_L - X_C)$

admittance: $Y = \dfrac{1}{Z}$

CIRCUIT LAWS

Kirchhoff's Voltage Law, KVL: The sum of the voltage rises equals the sum of the voltage drops in a closed loop.

Kirchhoff's Current Law, KCL: The sum of the currents into a node is zero.

Conservation of energy: Magnetic or electric energy cannot change instantaneously. For a capacitor, $v(0-) = v(0+)$; for an inductor, $i(0-) = i(0+)$.

Thevenin and Norton

v_{Th} = terminal open-circuit voltage

i_N = terminal short-circuit current

$Z_{Th} = \dfrac{v_{Th}}{i_N}$

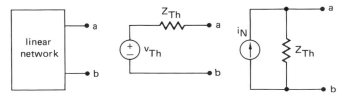

Superposition for linear networks means the contributions from each source can be determined independently and the results added.

$$i_3 = i_e + i_i$$

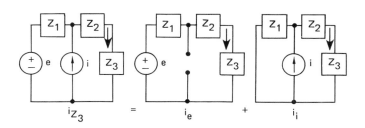

NETWORK REDUCTION

Current division through parallel elements is

$$I_1 = \dfrac{Z_2}{Z_1 + Z_2} I$$

$$I_2 = \dfrac{Z_1}{Z_1 + Z_2} I$$

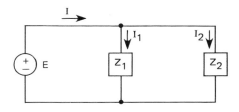

Voltage division across series elements is

$$V_1 = \dfrac{Z_1}{Z_1 + Z_2} E$$

$$V_2 = \dfrac{Z_2}{Z_1 + Z_2} E$$

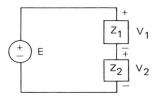

Converting pi (delta) to tee (wye) elements,

$$Z_1 = \dfrac{Z_{12} Z_{13}}{Z_{12} + Z_{13} + Z_{23}}$$

$$Z_2 = \dfrac{Z_{12} Z_{23}}{Z_{12} + Z_{13} + Z_{23}}$$

$$Z_3 = \dfrac{Z_{13} Z_{23}}{Z_{12} + Z_{13} + Z_{23}}$$

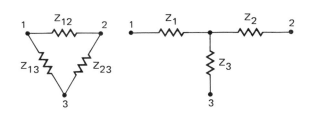

Converting tee (wye) to pi (delta) elements,

$$Z_{12} = \frac{Z_1 Z_2 + Z_1 Z_3 + Z_2 Z_3}{Z_3}$$

$$Z_{13} = \frac{Z_1 Z_2 + Z_1 Z_3 + Z_2 Z_3}{Z_2}$$

$$Z_{23} = \frac{Z_1 Z_2 + Z_1 Z_3 + Z_2 Z_3}{Z_1}$$

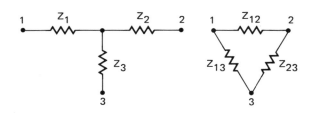

Z parameters:

$$V_1 = Z_{11} I_1 + Z_{12} I_2$$
$$V_2 = Z_{21} I_1 + Z_{22} I_2$$

Z_{11} = input impedance
Z_{12} = reverse transfer impedance
Z_{21} = forward transfer impedance
Z_{22} = output impedance

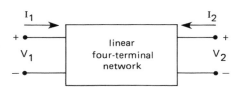

Y parameters:

$$I_1 = Y_{11} V_1 + Y_{12} V_2$$
$$I_2 = Y_{21} V_1 + Y_{22} V_2$$

Y_{11} = input admittance
Y_{12} = reverse transfer admittance
Y_{21} = forward transfer admittance
Y_{22} = output admittance

h parameters:

$$V_1 = h_{11} I_1 + h_{12} V_2$$
$$I_2 = h_{21} I_1 + h_{22} V_2$$

$h_i = h_{11}$ = input impedance
$h_r = h_{12}$ = reverse transfer voltage ratio
$h_f = h_{21}$ = forward transfer current ratio
$h_o = h_{22}$ = output admittance

COMBINING IMPEDANCES

Series

impedances: $Z = Z_1 + Z_2 + Z_3 + Z_4 + \cdots$

admittances: $\dfrac{1}{Y} = \dfrac{1}{Y_1} + \dfrac{1}{Y_2} + \dfrac{1}{Y_3} + \dfrac{1}{Y_4} + \cdots$

two admittances: $Y = \dfrac{Y_1 Y_2}{Y_1 + Y_2}$

Parallel

impedances: $\dfrac{1}{Z} = \dfrac{1}{Z_1} + \dfrac{1}{Z_2} + \dfrac{1}{Z_3} + \dfrac{1}{Z_4} + \cdots$

two impedances: $Z = \dfrac{Z_1 Z_2}{Z_1 + Z_2}$

admittances: $Y = Y_1 + Y_2 + Y_3 + Y_4 + \cdots$

MILLER'S THEOREM

Two voltages connected by an admittance can be converted to parallel admittances.

$$Y_a = Y_{ab}(1 - A)$$
$$Y_b = Y_{ab}\left(1 - \frac{1}{A}\right)$$

IDEAL TRANSFORMER

power converted to heat in transformation $= 0$

$$N_1 i_1 + N_2 i_2 + N_3 i_3 + \cdots = 0$$

$$\frac{v_1}{N_1} = \frac{v_2}{N_2} = \frac{v_3}{N_3} = \cdots$$

$$Z_1 = \frac{v_1}{i_1} = \left(\frac{N_1}{N_2}\right)^2 \frac{v_2}{i_2} = \left(\frac{N_1}{N_2}\right)^2 Z_2$$

$$L = \frac{N\phi}{i} \quad \phi = \text{flux}$$

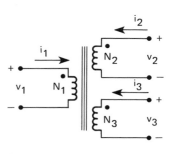

MUTUAL COUPLING

Dot terminal of coil is positive. Coils are linked by flux in the same direction.

$$v_1 = \frac{L_1 di_1}{dt} + \frac{M di_2}{dt}$$

$$v_2 = \frac{M di_1}{dt} + \frac{L_2 di_2}{dt}$$

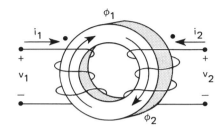

energy stored is $W = \frac{1}{2}L_1 i_1^2 + \frac{1}{2}L_2 i_2^2 + M i_1 i_2$

k = coefficient of coupling = $\dfrac{M}{\sqrt{L_1 L_2}}$

Coils are linked by flux in opposing directions.

$$v_1 = \frac{L_1 di_1}{dt} - \frac{M di_2}{dt}$$

$$v_2 = \frac{-M di_1}{dt} + \frac{L_2 di_2}{dt}$$

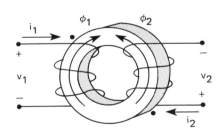

NETWORK TRANSFER FUNCTIONS

Parallel R-C

$$Y = \frac{1 + sCR}{R}$$

$$Z = \frac{R}{1 + sCR}$$

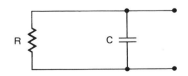

Series R-C

$$Y = \frac{sC}{1 + sCR}$$

$$Z = \frac{1 + sCR}{sC}$$

Series R-L

$$Y = \frac{1}{R\left(1 + \dfrac{sL}{R}\right)}$$

$$Z = R\left(1 + \frac{sL}{R}\right)$$

Series R-L-C

$$Y = \frac{sC}{1 + sCR + s^2 LC}$$

$$Z = \frac{1 + sCR + s^2 LC}{sC}$$

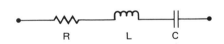

Parallel R-L-C

$$Y = \frac{1 + \dfrac{sL}{R} + s^2 LC}{sL}$$

$$Z = \frac{sL}{1 + \dfrac{sL}{R} + s^2 LC}$$

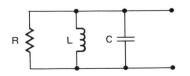

NETWORK ANALYSIS

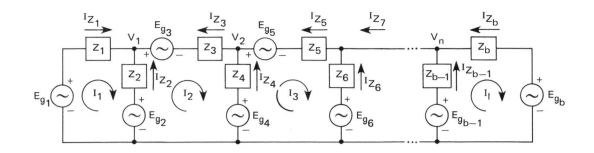

no. loops = no. branches − no. nodes
$$l = b - n$$

Branch Current Method

1: $I_{Z_1} + I_{Z_2} + I_{Z_3} = 0$
2: $-I_{Z_3} + I_{Z_4} + I_{Z_5} = 0$
. . . .
b: $-I_{Z_{b-2}} + I_{Z_{b-1}} + I_{Z_b} = 0$

1: $Z_1 I_{Z_1} - Z_2 I_{Z_2} + (0)I_{Z_3} = E_{g_1} - E_{g_2}$
2: $(0)I_{Z_1} + Z_2 I_{Z_2} - Z_3 I_{Z_3} - Z_4 I_{Z_4} = E_{g_2} - E_{g_3} - E_{g_4}$
. . . .
l: $Z_{(b-1)} I_{Z_{(b-1)}} - Z_b I_{Z_b} = E_{g_{(b-1)}} - E_{g_b}$

Loop Method

$$\begin{bmatrix} I_1 \\ I_2 \\ . \\ I_l \end{bmatrix} \begin{bmatrix} Z_{11} & -Z_{12} & -Z_{13} & \cdots & -Z_{1l} \\ -Z_{21} & Z_{22} & -Z_{23} & \cdots & -Z_{2l} \\ . & . & . & \cdots & . \\ -Z_{l1} & -Z_{l2} & -Z_{l3} & \cdots & Z_{ll} \end{bmatrix} = \begin{bmatrix} E_1 \\ E_2 \\ . \\ E_l \end{bmatrix}$$

$I_1, I_2, \ldots, I_l$ = loop currents
$E_1, E_2, \ldots, E_l$ = total voltage rises of each loop

$Z_{11}, Z_{22}, Z_{33}, \ldots, Z_{ll}$ = total impedance in loop l with other loops open circuited
Z_{12} = impedance common to loops 1 and 2 = Z_{21}
Z_{2l} = impedance common to loops 2 and l

Node Method

$$\begin{bmatrix} V_1 \\ V_2 \\ V_3 \\ . \\ V_n \end{bmatrix} \begin{bmatrix} Y_{11} & -Y_{12} & -Y_{13} & \cdots & -Y_{1n} \\ -Y_{21} & Y_{22} & -Y_{23} & \cdots & -Y_{2n} \\ -Y_{31} & -Y_{32} & Y_{33} & \cdots & -Y_{3n} \\ . & . & . & \cdots & . \\ -Y_{n1} & -Y_{n2} & -Y_{n3} & \cdots & Y_{nn} \end{bmatrix} = \begin{bmatrix} I_1 \\ I_2 \\ I_3 \\ . \\ I_n \end{bmatrix}$$

$V_1, V_2, V_3, \ldots, V_n$ = node voltages

$I_1, I_2, I_3, \ldots, I_n$ = sum of source currents into node n

$Y_{11}, Y_{22}, Y_{33}, \ldots, Y_{nn}$ = sum of admittances attached to node n

Y_{1n} = admittance from node 1 to node n = Y_{n1}

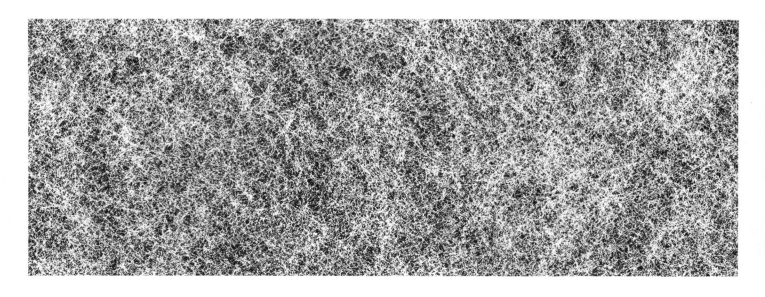

WAVEFORMS, POWER, AND MEASUREMENTS

AVERAGE VALUE

The average value of any waveform is the first term of the Fourier series.

$$A_0 = f_{\text{avg}} = \frac{1}{T}\int_{t_1}^{t_1+T} f(t)\,dt$$

$$= \frac{\text{positive area} - \text{negative area}}{T}$$

The average value of a sine or cosine wave over one cycle is zero. The average value of a sine or cosine wave over one-half cycle is $V_{\max}/\pi$.

Let the average value of a full-wave rectified sinusoid = $V_{\text{rec,avg}}$. Then with V_{thr} = threshold voltage and V_p = peak voltage,

$$\text{Let}\quad V_{\text{ratio}} = \frac{V_{\text{thr}}}{V_p}$$

$$\text{Let}\quad V_{\text{rad}} = \sqrt{1 - V_{\text{ratio}}^2}$$

$$\text{Let}\quad V_{\text{angle}} = \frac{2}{\pi}V_{\text{thr}}\arccos(V_{\text{ratio}})$$

$$V_{\text{rec,avg}} = \frac{2}{\pi}V_p\left(V_{\text{rad}} - V_{\text{angle}}\right)$$

$$\text{pulse train average value} = V_{\min} + (V_{\max} - V_{\min})\left(\frac{\Delta T}{T}\right)$$

$$\text{triangular wave average value} = \frac{1}{2}(V_{\max} + V_{\min})$$

$$\text{rectangular wave average value} = \left(\frac{V_{\max}}{T}\right)\Delta T$$

RMS VALUE

$$f_{\text{rms}}^2 = \frac{1}{T}\int_{t_1}^{t_1+T} f^2(t)\,dt$$

The rms value of a sine or cosine is $0.707 V_{\max}$. The rms value for half of a sine wave is $0.5 V_{\max}$.

Let $V_{\text{rec,rms}}$ = rms value of a full-wave rectified sinusoid. Then with V_{thr} = threshold voltage and V_p = peak voltage,

$$\text{Let}\quad V_{\text{ratio}} = \frac{V_{\text{thr}}}{V_p}$$

$$\text{Let}\quad V_{\text{rad}} = \sqrt{1 - V_{\text{ratio}}^2}$$

$$\text{Let}\quad V_{\text{angle}} = \frac{2}{\pi}V_{\text{thr}}\arccos(V_{\text{ratio}})$$

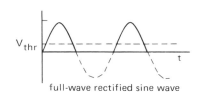
full-wave rectified sine wave

$$V_{\text{rms}}^2 = 0.5 V_p^2\left[(1 + 2V_{\text{ratio}}^2)V_{\text{angle}} - \frac{6}{\pi}V_{\text{ratio}}V_{\text{rad}}\right]$$

$$\text{pulse train rms voltage} = \sqrt{V_{\max}^2 - V_{\min}^2\left(\frac{\Delta T}{T}\right) + V_{\min}^2}$$

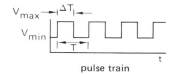
pulse train

$$\text{triangular wave rms voltage} = V_{\text{tr,rms}}$$

triangular

$$V_{\text{tr,rms}} = \sqrt{(V_{\max}^2 + V_{\min}^2 + V_{\max}V_{\min})\left(\frac{|t_a| + t_b}{3T}\right)}$$

$$\text{rectangular wave rms voltage} = V_{\max}\sqrt{\frac{\Delta T}{T}}$$

rectangular

$$\text{Fourier series} = \sqrt{\begin{array}{l}\text{square of DC value plus the sum of}\\\text{the squares of the rms values of}\\\text{each harmonic}\end{array}}$$

$$= \sqrt{(A_0)^2 + \frac{1}{2}\sum(a_n^2 + b_n^2)}$$

SPECTRA

A spectrum is a plot of voltage, current, or power magnitude for each Fourier harmonic, including the average value (zero-frequency term). The spectrum is usually given in rms values for each harmonic, n.

For power,

$$P(0) = (A_0)^2$$

$$P(1) = \frac{1}{2}(a_1^2 + b_2^2)$$

$$. = \ldots..$$

$$P(n) = \frac{1}{2}(a_n^2 + b_n^2)$$

For voltage,

$$V(0) = \sqrt{P(0)}$$

$$V(1) = \sqrt{P(1)}$$

$$. = \ldots..$$

$$V(n) = \sqrt{P(n)}$$

PROFESSIONAL PUBLICATIONS, INC. • Belmont, CA

VOLTMETERS

d'Arsonval movement measures the average value of a wave.

sensitivity of movement $= \dfrac{1}{I_{fs}} = \Omega/V$

I_{fs} = full scale current value
typical values = 1000; 5000; 10,000; 20,000 in Ω/V

$$\frac{R_{coil} + R_{ext}}{V_{fs}} = \frac{1}{I_{fs}}$$

V_{fs} = full scale voltage
R_{ext} = external resistor

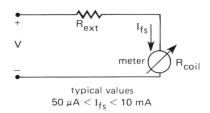

typical values
$50\ \mu A < I_{fs} < 10\ mA$

AMMETERS

$$I_{fs} = \frac{V_{fs}}{R_{shunt}} + \frac{V_{fs}}{R_{coil} + R_{swamp}}$$

R_{shunt} = shunt resistor
R_{swamp} = swamping resistor

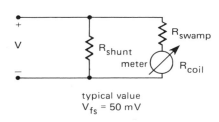

typical value
$V_{fs} = 50\ mV$

OPERATIONAL AMPLIFIER METERS

half-wave rectifier meter

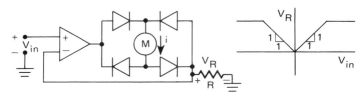

full-wave rectifier meter

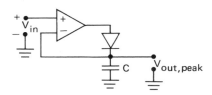

peak voltage detector

AVERAGE POWER

apparent power = resistive power + reactive power

$$VA = W + VA\ reactive$$
$$S = P + jQ = VI^*$$
I^* = complex conjugate of I
$$S^2 = P^2 + Q^2$$

power factor $= \text{cosine}\ \theta = \dfrac{|P|}{|S|}$

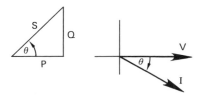

inductive power triangle
lagging (+) power factor

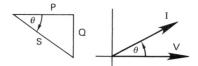

capacitive power triangle
leading (−) power factor

POWER FACTOR CORRECTION

kVAR capacitors required $= P(\tan\theta_1 - \tan\theta_2)$

$$\text{capacitive reactance} = X_C = \frac{V^2}{Q_1 - P(\tan\theta_2)}$$

$$= \frac{V^2}{Q_1 - \dfrac{P\sqrt{1-\text{pf}^2}}{\text{pf}}}$$

$\text{pf} = \cos\theta_2$

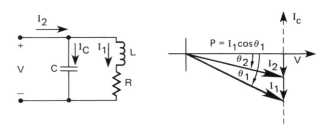

TIME AND FREQUENCY RESPONSE

FIRST-ORDER TRANSIENTS

A first-order circuit contains one energy storage element.

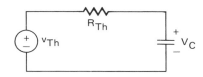

R_{Th} = Thevenin resistance
v_{Th} = Thevenin voltage
$= R_{Th} C \dfrac{dV_C}{dt} + V_C$

In the time domain t,

$$V_C(t) = Ae^{-t/R_{Th}C} + V_{C,ss}$$

A = constant

In the complex domain s,

$$V_C(s) = \dfrac{V_{Th}(s) + R_{Th}CV_C(0)}{sCR_{Th} + 1}$$

$V_{C,ss}$ = limit $V_C(s)$ as $s \to 0$

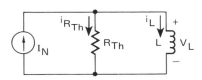

R_{Th} = Thevenin resistance
I_N = Norton current

$$i_{R_{Th}} = \dfrac{V_L}{R_{Th}} = \dfrac{L\left(\dfrac{di_L}{dt}\right)}{R_{Th}}$$

$$I_N = \left(\dfrac{L}{R_{Th}}\right)\dfrac{di_L}{dt} + i_L$$

In the time domain t,

$$i_L(t) = Ae^{-t/(L/R_{Th})} + I_{L,ss}$$

A = constant

In the complex domain s,

$$I_L(s) = \dfrac{R_{Th}I_N(s) + Li_L(0)}{sL + R_{Th}}$$

$i_{L,ss}$ = limit $I_L(s)$ as $s \to 0$

SECOND-ORDER TRANSIENTS

A second-order circuit contains two energy storage elements and has a differential equation of the form

$$f(t) = a\dfrac{d^2q}{dt^2} + \dfrac{dq}{dt} + cq$$

At time $t = 0^+$,

$q(0^+)$ = initial condition

$q'(0^+)$ = value of $\dfrac{dq}{dt}$

$q''(0^+)$ = value of $\dfrac{d^2q}{dt^2}$

q_{ss} = steady-state value of q

Case I: Overdamped

$$b^2 > 4ac$$

Overdamped transients decay without oscillations.

$$q(t) = Ae^{S_1(t)} + Be^{S_2(t)} + q_{ss}$$

$$S_1 = \dfrac{-b + \sqrt{b^2 - 4ac}}{2a}$$

$$S_2 = \dfrac{-b - \sqrt{b^2 - 4ac}}{2a}$$

$$A = \left(\dfrac{1}{2}\right)\left(1 + \dfrac{b}{\sqrt{b^2 - 4ac}}\right)[q(0^+) - q_{ss}]$$
$$+ \left(\dfrac{a}{\sqrt{b^2 - 4ac}}\right)[q'(0^+) - q'_{ss}]$$

$$B = \left(\dfrac{1}{2}\right)\left(1 - \dfrac{b}{\sqrt{b^2 - 4ac}}\right)[q(0^+) - q_{ss}]$$
$$- \left(\dfrac{a}{\sqrt{b^2 - 4ac}}\right)[q'(0^+) - q'_{ss}]$$

Case II: Critically Damped

$$b^2 = 4ac$$

Critical damping is the fastest decay without oscillations.

$$q(t) = Ae^{st} + Bte^{st} + q_{ss}$$

$$S = \dfrac{-b}{2a}$$

$$A = q(0^+) - q_{ss}$$

$$B = q'(0^+) - q'_{ss} + \dfrac{b}{2a}A$$

Case III: Underdamped

$$b^2 < 4ac$$

Underdamped transients decay with oscillations.

$$q(t) = Ae^{-\alpha t}\cos\beta t + Be^{-\alpha t}\sin\beta t + q_{ss}$$
$$\alpha = \frac{b}{2a}$$
$$\beta = \frac{\sqrt{4ac - b^2}}{2a}$$
$$A = q(0^+) - q_{ss}$$
$$B = \left(\frac{\alpha}{\beta}\right)A + \left(\frac{1}{\beta}\right)[q'(0^+) - q'_{ss}]$$

LAPLACE ANALYSIS

Include initial conditions as circuit elements. Solve using circuit reduction techniques or circuit analysis.

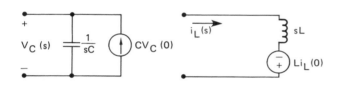

DECIBEL

$$1 \text{ bel} = \log_{10}\left(\frac{p}{p_{\text{reference}}}\right)$$
$$\frac{1}{10}\text{ bel} = 1\text{ dB}, \quad p = \text{power}, \quad V = \text{voltage}$$
$$\text{dB} = 10\log_{10}\left(\frac{p}{p_{\text{reference}}}\right)$$
$$\text{dB} = 20\log_{10}\left(\frac{V}{V_{\text{reference}}}\right)$$

TRANSFER FUNCTIONS

transfer function = output/input (real or complex)

BODE DIAGRAMS

A Bode diagram is a plot of transfer function gain in dB and phase angle versus logarithm to the base 10 of frequency. A decade is a frequency 10 times as great as another frequency. An octave is a frequency two times as great as another frequency.

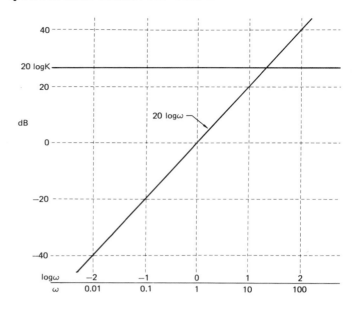

$$T = s = j\omega$$
$$|T| = |s| = \omega$$
$$K = \text{constant}$$

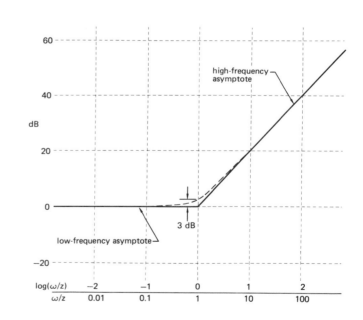

$$|T| = 20\log\left|1 + \frac{s}{z}\right|$$

When $|s| = z$,

$$|T| = 20\log\left(\sqrt{2}\right) = 3.01\text{ dB}$$

TIME AND FREQUENCY RESPONSE

First-Order Filters

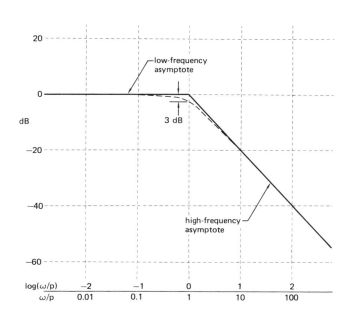

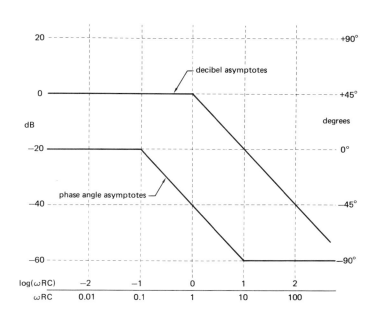

$$|T| = -20 \log \left|1 + \frac{s}{p}\right|$$

When $|s| = p$,

$$|T| = -20 \log\left(\sqrt{2}\right) = -3.01 \text{ dB}$$

Low-Pass Filters

$$T_v(s) = \frac{1}{1 + sCR}$$

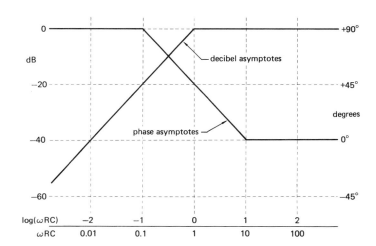

SECOND-ORDER EQUATION FORMS

$$s^2 + 2\zeta\omega_n s + \omega_n^2$$

$$s^2 + \left(\frac{\omega_o}{Q}\right)s + \omega_o^2$$

$$(s + \alpha)^2 + \beta^2$$

$$\omega_0^2 = \omega_n^2 = \alpha^2 + \beta^2$$

$$\zeta = \frac{1}{2Q}$$

$$\alpha = \frac{\omega_0}{2Q}$$

Quality factor, Q, is a measure of circuit selectivity. In general, $Q = \omega L/R$.

High-Pass Filters

$$T_v(s) = \frac{sCR}{1 + sCR}$$

SECOND-ORDER SYSTEMS

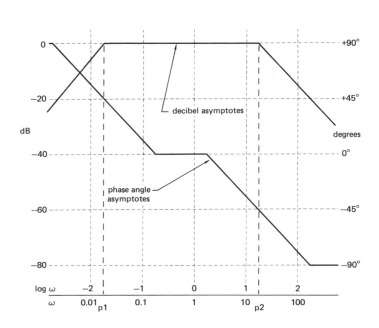

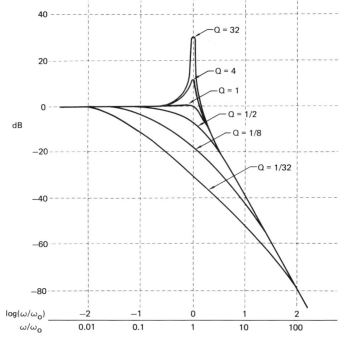

Low-Pass Filters

$$G_{\text{LP}}(s) = \cfrac{1}{1 + \cfrac{s}{Q\omega_o} + \cfrac{s^2}{\omega_o^2}}$$

Band-Pass Filters

$$H(s) = \cfrac{\cfrac{s}{\omega_o}}{\left(1 + \cfrac{s}{p_1}\right)\left(1 + \cfrac{s}{p_2}\right)} = \cfrac{\cfrac{s}{\omega_o}}{1 + \cfrac{s}{Q\omega_o} + \cfrac{s^2}{\omega_o^2}}$$

$$\text{BW} = \text{bandwidth} = \omega_2 - \omega_1 = \frac{\omega_o}{Q}$$

Half-power frequencies are

$$\omega_1 = \frac{\omega_o}{2Q(\sqrt{1 + 4Q^2} - 1)}$$

$$\omega_2 = \frac{\omega_o}{2Q(\sqrt{1 + 4Q^2} + 1)}$$

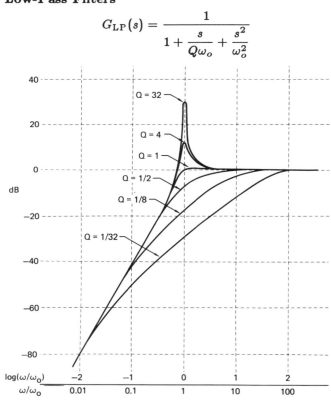

High-Pass Filters

$$G_{\text{HP}}(s) = \cfrac{\cfrac{s^2}{\omega_o^2}}{1 + \cfrac{s}{Q\omega_o} + \cfrac{s^2}{\omega_o^2}}$$

TIME AND FREQUENCY RESPONSE

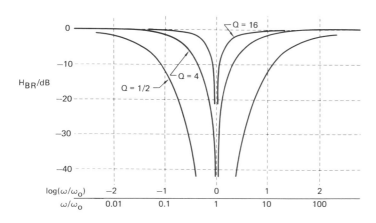

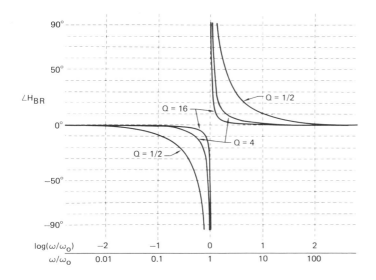

Band-Reject Filters

$$H_{\text{BR}}(s) = \frac{\dfrac{s^2}{\omega_o^2} + 1}{1 + \dfrac{s}{Q\omega_o} + \dfrac{s^2}{\omega_o^2}}$$

$\text{BW} = \text{bandwidth} = \omega_2 - \omega_1 = \dfrac{\omega_o}{Q}$

Half-power frequencies are

$\omega_1 = \dfrac{\omega_o}{2Q}\left(\sqrt{4Q^2 + 1} - 1\right)$

$\omega_2 = \dfrac{\omega_o}{2Q}\left(\sqrt{4Q^2 + 1} + 1\right)$

POWER SYSTEMS

POSITIVE PHASE SEQUENCE

Positive phase sequence phasors exist as three equal magnitude phasors rotating counterclockwise in phase sequence A, B, C. The phasors are 120 electrical degrees apart and their sum is zero.

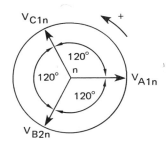

NEGATIVE PHASE SEQUENCE

Negative phase sequence phasors exist as three equal magnitude phasors rotating counterclockwise in phase sequence A, C, B. The phasors are 120 electrical degrees apart and their sum is zero.

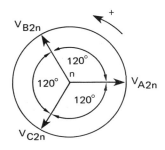

ZERO PHASE SEQUENCE

Zero phase sequence phasors exist as three equal magnitude phasors coincident in phase sequence.

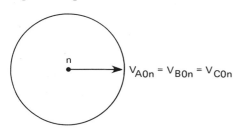

SYMMETRICAL COMPONENTS

Any three-phase system of voltages (or currents) can be resolved into the sum of positive, negative, and zero phase sequence components. The following formulas are referenced to neutral.

$$\text{operator } a = 1\underline{/120°} = -0.5 + j0.866$$
$$a^2 = 1\underline{/240°} = -0.5 - j0.866$$
$$a^3 = 1$$
$$1 + a + a^2 = 0$$

$$V_{a0} = \frac{1}{3}(V_a + V_b + V_c)$$
$$V_{a1} = \frac{1}{3}(V_a + aV_b + a^2V_c)$$
$$V_{a2} = \frac{1}{3}(V_a + a^2V_b + aV_c)$$

$$V_a = V_{a0} + V_{a1} + V_{a2}$$
$$V_b = V_{a0} + a^2V_{a1} + aV_{a2}$$
$$V_c = V_{a0} + aV_{a1} + a^2V_{a2}$$

BALANCE SOURCES AND LOADS

Either positive sequence only or negative sequence only voltages and currents exist on balanced systems. The phasor sum of line-to-line voltages is zero. The phasor sum of line currents is zero. The phase powers are all equal and are one-third of the total power. There is no neutral current flow if a neutral wire is used.

$$\text{total power} = \sqrt{3}\, V_l\, I_l \cos\theta$$

a = turns ratio between secondary and primary per phase

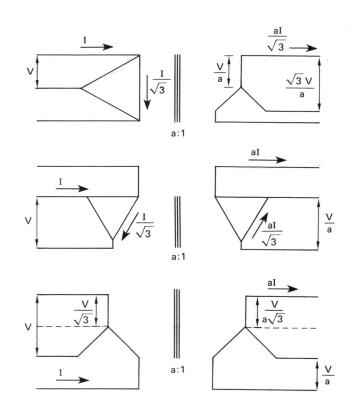

UNBALANCED SOURCES AND LOADS

The phasor sum of line-to-line voltages is zero. The unbalanced voltage and currents can be resolved into the sum of positive, negative, and zero sequence voltages and currents.

POWER SYSTEMS

The total average power is equal to the sum of the positive, negative, and zero phase sequence powers. Voltages and currents are referenced to neutral.

$$\begin{array}{llll} V_a\angle\alpha_a & I_a\angle\beta_a & V_{a0}\angle\alpha_0 & I_{a0}\angle\beta_0 \\ V_b\angle\alpha_b & I_b\angle\beta_b & V_{a1}\angle\alpha_{a1} & I_{a1}\angle\beta_{a1} \\ V_c\angle\alpha_c & I_c\angle\beta_c & V_{a2}\angle\alpha_{a2} & I_{a2}\angle\beta_{a2} \end{array}$$

$$\begin{aligned} \text{average power} &= V_a I_a \cos(\beta_a - \alpha_a) \\ &\quad + V_b I_b \cos(\beta_b - \alpha_b) \\ &\quad + V_c I_c \cos(\beta_c - \alpha_c) \\ &= 3V_{a0}I_{a0}\cos(\beta_0 - \alpha_0) \\ &\quad + 3V_{a1}I_{a1}\cos(\beta_{a1} - \alpha_{a1}) \\ &\quad + 3V_{a2}I_{a2}\cos(\beta_{a2} - \alpha_{a2}) \end{aligned}$$

UNBALANCED THREE-WIRE SYSTEMS

The phasor sum of line currents is zero. No zero phase sequence currents exist on the lines.

$$I_a + I_b + I_c = 0$$
$$I_a = I_{a1} + I_{a2}$$
$$I_b = a^2 I_{a1} + a I_{a2}$$
$$I_c = a I_{a1} + a^2 I_{a2}$$

The phasor sum of phase currents for a wye system is zero. The phasor sum of phase currents for a delta system is not zero and is equal to the sum of the zero sequence currents circulating in the delta phases. The phasor sum of line-to-line voltages is zero. The phasor sum of phase voltages is not zero.

UNBALANCED FOUR-WIRE SYSTEMS

The phasor sum of line currents is not zero.

$$I_a = I_{a0} + I_{a1} + I_{a2}$$
$$I_b = I_{a0} + a^2 I_{a1} + a I_{a2}$$
$$I_c = I_{a0} + a I_{a1} + a^2 I_{a2}$$

A zero sequence current flows in the neutral wire. The phasor sum of the line-to-line voltages is zero. The phasor sum of the phase voltages is not zero.

$$V_{an} = V_{an0} + V_{an1} + V_{an2}$$
$$V_{bn} = V_{an0} + a^2 V_{an1} + a V_{an2}$$
$$V_{cn} = V_{an0} + a V_{an1} + a^2 V_{an2}$$

POWER TRANSFORMERS

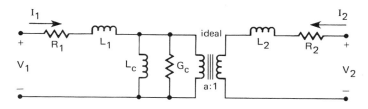

equivalent circuit model

OPEN-CIRCUIT TEST

The open-circuit test measures the core losses.

$$V_{oc} = \text{rated primary voltage}$$
$$P_{oc} = V_{oc}^2 G_c$$
$$Y_c = G_c - jB_c$$
$$|B_c| = \frac{\sqrt{S_{oc}^2 - P_{oc}^2}}{V_{oc}^2}$$

SHORT-CIRCUIT TEST

The short-circuit test measures the primary and secondary impedances.

$$I_{sc} = \text{rated secondary current}$$
$$Z_{sc} = R_1 + jX_1 + a^2(R_2 + jX_2)$$

For most transformers,

$$R_1 = \frac{P_{sc}}{2I_{sc}^2} = a^2 R_2$$

$$X_1 = \frac{Q_{sc}}{2I_{sc}^2} = a^2 X_2$$

ABCD PARAMETERS

$$V_{in} = AV_{out} - BI_{out}$$
$$I_{in} = CV_{out} - DI_{out}$$

For the transformer equivalent circuit,

$$a = \text{primary-to-secondary turns ratio}$$

$$Z_1 = R_1 + jX_1$$
$$Z_2 = R_2 + jX_2$$
$$Y_c = G_c - jB_c$$

$$V_1 = (1 + Z_1 Y_c)aV_2 - \frac{[Z_1 + a^2 Z_2(1 + Z_1 Y_c)]I_2}{a}$$

$$I_1 = aY_c V_2 - \frac{(1 + a^2 Z_2 Y_c)I_2}{a}$$

PER-UNIT SYSTEM

$$\text{per unit} = \text{P.U.} = \frac{\text{actual}}{\text{base}}$$

$$= \frac{\text{percent}}{100}$$

Bases chosen are usually kVA and kV of a transformer. Per units of the same bases can be combined.

ϕ = phase

ln = line-to-neutral

V_{ll} = line-to-line = $\sqrt{3}\, V_{ln}$

$\text{kW}_{\text{base},\phi} = \text{kVA}_{\text{base},\phi} = \text{kVAR}_{\text{base},\phi}$

$$I_{\text{base}} = \frac{\text{kVA}_{\text{base},\phi}}{\text{kV}_{\text{base},ln}}$$

$$Z_{\text{base}} = \frac{V_{\text{base},ln}}{I_{\text{base}}}$$

$$= \frac{V_{\text{base},ln}^2}{1000\,\text{kVA}_{\text{base},\phi}}$$

$$Z_{\text{P.U.}} = \frac{\Omega \times \text{kVA}_{\text{base},\phi}}{1000\,\text{kV}_{\text{base},ln}^2}$$

Base Conversions for Per Unit

$$I_{\text{new P.U.}} = I_{\text{old P.U.}} \left(\frac{\text{kVA}_{\text{old base}}}{\text{kVA}_{\text{new base}}}\right)\left(\frac{\text{kV}_{\text{new base}}}{\text{kV}_{\text{old base}}}\right)$$

$$Z_{\text{new P.U.}} = Z_{\text{old P.U.}} \left(\frac{\text{kVA}_{\text{new base}}}{\text{kVA}_{\text{old base}}}\right)\left(\frac{\text{kV}_{\text{old base}}}{\text{kV}_{\text{new base}}}\right)^2$$

AUTOTRANSFORMERS

One line of the primary and secondary is common. Solve the same as isolated primary transformers. Autotransformers are generally used to boost to a higher voltage.

FAULTS

In general, a short circuit must be solved using symmetrical components. The load current is much smaller than the fault current and is ignored. The magnitude of the fault current is dependent on the kVA of the supplying system and the time on the sine wave of fault initiation (maximum at zero voltage).

ϕn = phase to neutral

Z_0 = zero sequence reactance from fault to neutral

Z_1 = positive sequence reactance from fault to neutral

Z_2 = negative sequence reactance from fault to neutral

Single Line-to-Neutral Fault

$$I_{\text{fault}} = \frac{3V_{\phi n}}{Z_0 + Z_1 + Z_2}$$

Double Line-to-Neutral Fault

$$I_{\text{fault}} = \frac{3V_{\phi n} Z_2}{Z_1 Z_2 + Z_0(Z_1 + Z_2)}$$

Line-to-Line Fault

$$I_{\text{fault}} = \frac{V_{\phi n}\sqrt{3}}{Z_1 + Z_2}$$

Three-Phase (Bolted) Fault

$$I_{\text{fault}} = \frac{V_{\phi n}}{Z_1}$$

TRANSMISSION LINES

FREQUENCY AND WAVELENGTH

The wavelength of a transverse electromagnetic wave is given by

$$\text{wavelength} = \frac{\text{velocity}}{\text{frequency}}$$

$$\lambda = \frac{v}{f}$$

POWER TRANSMISSION LINES

Use the following correction factors for other than 1-foot spacing. D, GMR, and r are in feet.

$$K_L = 1 + \frac{\ln D}{\ln\left(\frac{1}{\text{GMR}}\right)}$$

$X_L = X_L$ from table times K_L

$$K_C = 1 + \frac{\ln D}{\ln\left(\frac{1}{r}\right)}$$

$X_C = X_C$ from table times K_C

Electrical Characteristics of the Multilayer, Steel-Reinforced, Bare Aluminum Conductors

Code Name	Aluminum Area (c.m.)	Strands (Al/St)	O.D. (in)	R_{DC} 20°C (mΩ/ft)	R_{AC}, 60 Hz 20°C (Ω/mi)	R_{AC}, 60 Hz 50°C (Ω/mi)	GMR (ft)	1-ft spacing reactance per conductor @ 60 Hz X_L (Ω/mi)	1-ft spacing reactance per conductor @ 60 Hz X_C (MΩ-mi)
waxwing	266,800	18/1	0.609	0.0646	0.3488	0.3831	0.0198	0.476	0.1090
partridge	266,800	26/7	0.642	0.0640	0.3452	0.3792	0.0217	0.465	0.1074
ostrich	300,000	26/7	0.680	0.0569	0.3070	0.3372	0.0229	0.458	0.1057
merlin	336,400	18/1	0.684	0.0512	0.2767	0.3037	0.0222	0.462	0.1055
linnet	336,400	26/7	0.721	0.0507	0.2737	0.3006	0.0243	0.451	0.1040
oriole	336,400	30/7	0.741	0.0504	0.2719	0.2987	0.0255	0.445	0.1032
chickadee	397,500	18/1	0.743	0.0433	0.2342	0.2572	0.0241	0.452	0.1031
ibis	397,500	26/7	0.783	0.0430	0.2323	0.2551	0.0264	0.441	0.1015
pelican	477,000	18/1	0.814	0.0361	0.1957	0.2148	0.0264	0.441	0.1004
flicker	477,000	24/7	0.846	0.0359	0.1943	0.2134	0.0284	0.432	0.0992
hawk	477,000	26/7	0.858	0.0357	0.1931	0.2120	0.0289	0.430	0.0988
hen	477,000	30/7	0.883	0.0355	0.1919	0.2107	0.0304	0.424	0.0980
osprey	556,500	18/1	0.879	0.0309	0.1679	0.1843	0.0284	0.432	0.0981
parakeet	556,500	24/7	0.914	0.0308	0.1669	0.1832	0.0306	0.423	0.0969
dove	556,500	26/7	0.927	0.0307	0.1663	0.1826	0.0314	0.420	0.0965
rook	636,000	24/7	0.977	0.0269	0.1461	0.1603	0.0327	0.415	0.0950
grosbeak	636,000	26/7	0.990	0.0268	0.1454	0.1596	0.0335	0.412	0.0946
drake	795,000	26/7	1.108	0.0215	0.1172	0.1284	0.0373	0.399	0.0912
tern	795,000	45/7	1.063	0.0217	0.1188	0.1302	0.0352	0.406	0.0925
rail	954,000	45/7	1.165	0.0181	0.0977	0.1092	0.0386	0.395	0.0897
cardinal	954,000	54/7	1.196	0.0180	0.0988	0.1082	0.0402	0.390	0.0890
ortolan	1,033,500	45/7	1.213	0.0167	0.0924	0.1011	0.0402	0.390	0.0885
bluejay	1,113,000	45/7	1.259	0.0155	0.0861	0.0941	0.0415	0.386	0.0874
finch	1,113,000	54/19	1.293	0.0155	0.0856	0.0937	0.0436	0.380	0.0866
bittern	1,272,000	45/7	1.382	0.0136	0.0762	0.0832	0.0444	0.378	0.0855
bobolink	1,431,000	45/7	1.427	0.0121	0.0684	0.0746	0.0470	0.371	0.0837
plover	1,431,000	54/19	1.465	0.0120	0.0673	0.0735	0.0494	0.365	0.0829
lapwing	1,590,000	45/7	1.502	0.0109	0.0623	0.0678	0.0498	0.364	0.0822
falcon	1,590,000	54/19	1.545	0.0108	0.0612	0.0667	0.0523	0.358	0.0814
bluebird	2,156,000	84/19	1.762	0.0080	0.0476	0.0515	0.0586	0.344	0.0776

Reproduced by permission from *Aluminum Electrical Conductor Handbook*, published by the Aluminum Association, 1989.

DISTRIBUTED CONSTANTS LINES

Constants are given in Ω per unit length.

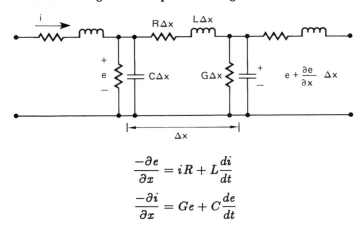

$$\frac{-\partial e}{\partial x} = iR + L\frac{di}{dt}$$

$$\frac{-\partial i}{\partial x} = Ge + C\frac{de}{dt}$$

LOSSLESS LINES

Lossless lines are lines where $R = G = 0$. The velocity of a wave on a line $= 1/\sqrt{LC}$ unit length per second.

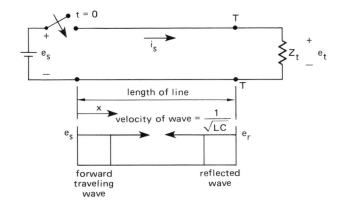

LOSSLESS LINES WITH REFLECTIONS

The sum of the sending and reflected voltage wave is the total line voltage at a given point.

$$e_t = e_s + e_r$$

z_o = line characteristic impedance

ρ = reflection coefficient

Z_t = terminating impedance of line

$$Z_t = \frac{e_t}{i_t} = \frac{e_s + e_r}{i_s + i_r}$$

$$= \frac{e_s + e_r}{\frac{e_s}{z_o} - \frac{e_r}{z_o}}$$

$$\rho = \frac{e_r}{e_s} = \frac{Z_t - z_o}{Z_t + z_o}$$

When $Z_t = z_o$, the line has no reflections and $\rho = 0$.

AC STEADY-STATE LINES

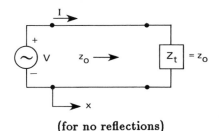

(for no reflections)

$$z = R + j\omega L$$
$$y = G + j\omega C$$
$$\frac{dV}{dx} = -zI$$
$$\frac{dI}{dx} = -yV$$
$$\frac{d^2V}{dx^2} = yZV$$
$$V = V_1 e^{-\gamma x} + V_2 e^{+\gamma x}$$

x = length along line

z_o = characteristic impedance of line $= \sqrt{\dfrac{z}{y}}$

V_1 is a forward traveling voltage.
V_2 is a reflected traveling voltage.

$$z_o = \sqrt{\frac{R + j\omega L}{G + j\omega C}}$$

$$Iz_o = V_1 e^{-\gamma x} + V_2 e^{+\gamma x}$$

γ = propagation constant $= \sqrt{yz}$
$\quad = \sqrt{(R + j\omega L)(G + j\omega C)}$

real $\gamma = \alpha$ = attenuation constant = neper/mi

imaginary $\gamma = \beta$ = phase constant = rad/mi

LONG LINES

For length $>$ 150 miles and 60 Hz frequency, use AC steady-state line equations.

MEDIUM LENGTH LINES

For length 50–150 miles and 60 Hz frequency, use tee or pi approximation.

Tee Approximation

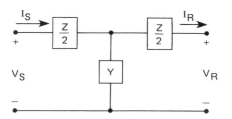

$$V_S = AV_R + BI_R$$
$$I_S = CV_R + DI_R$$
$$A = 1 + \frac{YZ}{2}$$
$$B = Z\left(1 + \frac{YZ}{4}\right)$$
$$C = Y$$
$$D = 1 + \frac{YZ}{2}$$

Pi Approximation

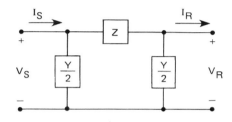

$$V_S = AV_R + BI_R$$
$$I_S = CV_R + DI_R$$
$$A = 1 + \frac{YZ}{2}$$
$$B = Z$$
$$C = Y\left(1 + \frac{YZ}{4}\right)$$
$$D = 1 + \frac{YZ}{2}$$

SHORT LINES

At 60 Hz and lengths less than 50 miles, the shunt reactances are ignored.

$$I_S = I_R$$
$$V_S = V_R + I_S(R + jX)$$

$$\text{percent line voltage regulation} = \frac{|V_{R,\text{no load}}| - |V_{R,\text{full load}}|}{|V_{R,\text{full load}}|} \times 100\%$$

HIGH-FREQUENCY LINES

Assumptions are $R = 0$, and $G_C = $ infinity.

$$z_o = \sqrt{\frac{L}{C}}$$
$$z = j\omega L$$
$$y = j\omega C$$

$\gamma = $ propagation constant $= j\omega\sqrt{LC}$

$\beta = $ phase constant $= \omega\sqrt{LC}$

$\alpha = $ attenuation constant $= 0$

$V_1 = $ forward traveling wave voltage

$V_2 = $ reflected traveling wave voltage

$V_R = V_1 + V_2$

$V_S = V_1[(1 + \rho)\cos\beta l + j(1 - \rho)\sin\beta l]$

$I_S = \dfrac{V_1}{z_o}[(1 - \rho)\cos\beta l + j(1 + \rho)\sin\beta l]$

$Z_i = $ input impedance $= \dfrac{V_S}{I_S}$

VSWR $= $ voltage standing wave ratio $= \dfrac{1 + \rho}{1 - \rho}$

$$Z_{\max} = z_o \text{VSWR}$$

$$Z_{\min} = \frac{z_o}{\text{VSWR}}$$

SMITH CHART

Smith chart distances can be used to find ρ and VSWR.

$$\rho = \frac{\overline{OZ}}{\overline{OA}}$$

$$\text{VSWR} = \frac{\overline{AB}}{\overline{BC}}$$

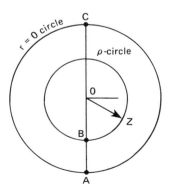

Smith chart

Use normalized impedance $Z_n = \dfrac{R + jX}{z_o}$.

ROTATING MACHINES

FARADAY'S LAW

e = induced voltage in V = $\dfrac{-N d\phi}{dt}$

N = number of turns

ϕ = flux in Wb = BA

A = area normal to flux in m^2

B = flux density in Wb/m^2

DC GENERATORS

E_g = generated voltage in V = $K_v \phi \Omega$

Ω = mechanical angular velocity in rad/s

ϕ = flux in Wb

K_v = voltage constant

$P_{\text{shaft,in}}$ − mechanical losses = $E_g I_a = T_{\text{air gap}} \Omega$

$T_{\text{air gap}}$ = air-gap torque in N·m = $K_t \phi I_a$

K_t = torque constant = K_v in SI system

I_a = armature current in A

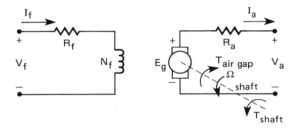

DC MOTORS

Same as DC generator except reverse I_a and $T_{\text{air gap}}$.

$P_{\text{shaft,out}}$ + mechanical losses = $T_{\text{air gap}} \Omega = E_g I_a$

SELF-EXCITED GENERATORS

$$E_g = K_v \phi \Omega = V_a + R_a I_L + R_a I_f$$

a = long shunt
b = short shunt

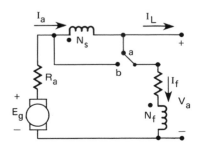

MMF = field MMF + armature MMF

MMF = $N_f I_f + N_s I_L$ for short shunt

OPS = slope of magnetization curve at the operating point

$$\dfrac{\partial E_g}{\partial I_L} = \left(\dfrac{N_s}{N_f}\right)\left(\dfrac{E_n}{I_n}\right)(\text{OPS}) = R_a$$

$$I_f = I_n \text{MMF}_n - \left(\dfrac{N_s}{N_f}\right) I_L$$

$$\dfrac{E_g}{E_n} = \dfrac{V_a + R_a I_L \left(1 - \dfrac{N_s}{N_f}\right)}{E_n} + R_a \left(\dfrac{I_n}{E_n}\right) \text{MMF}_n$$

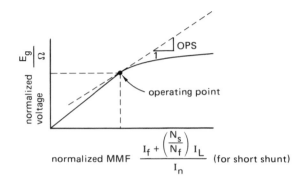

SEPARATELY EXCITED DC MOTORS

$$V_a = I_a R_a + K_v \phi \Omega$$

$$T = K_t \phi I_a = \left(\dfrac{K_t \phi}{R_a}\right)(V_a - K_v \phi \Omega)$$

$$= J \dfrac{d\Omega}{dt} + f_v \Omega$$

J = mass moment of inertia in N·m·s^2/rad

f_v = coefficient of viscous friction in N·m·s/rad

K_t and K_v = constants

Ω = mechanical angular velocity in rad/s

SERIES MOTORS

$$V_a = K_v \phi \Omega + I_a(R_a + R_s)$$

$$T = K_t \phi I_a$$

$$\text{MMF} = N_s I_a$$

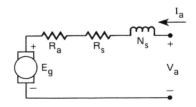

ROTATING MACHINES

SHUNT MOTORS

$$P_{\text{in}} = V_a(I_a + I_f) = V_a I_a + \frac{V_a^2}{R_f}$$

$$V_a I_a = E_g I_a + I_a^2 R_a$$

$$E_g I_a = T\Omega = \text{rotational losses} + T_L \Omega$$

For constant speed at an operating point on the magnetization curve,

$$\frac{N_s}{N_f} = \frac{(R_a + R_s)\left(\frac{I_n}{E_n}\right)}{\text{OPS}}$$

OPS = slope of curve at operating point

Cumulative Compounding Motors

$$\text{MMF} = N_f I_f + N_s I_a$$

Differential Compounding Motors

$$\text{MMF} = N_f I_f - N_s I_a$$

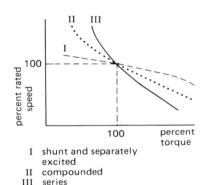

I   shunt and separately excited
II compounded
III series

DC MACHINE LOSSES

electrical power = field power + armature power
                    + air gap power

air gap power = load power + mechanical losses (motor)
              = load power − mechanical losses (generator)

STARTING DC MACHINES

Series Resistance

A series resistance limits starting current and is reduced as E_g increases with speed.

$$\frac{V_a - E_g}{R_a + R_{\text{series}}} = I_a$$

R_{series} = series resistance

Reduced Voltage

Armature voltage is supplied from a variable source such as another DC generator or solid-state device.

SPEED CONTROL OF DC MACHINES

$$\Omega = \frac{V_a - I_a R_a}{K_v \phi}$$

Assuming a linear magnetization curve,

$$K_t \phi = \frac{E_g}{\Omega} = K_f I_f$$

$$\Omega = \frac{V_a - I_a R_a}{K_f I_f}$$

$$S = \text{speed ratio} = \frac{\Omega_1}{\Omega_2}$$
$$= \left(\frac{I_{f2}}{I_{f1}}\right)\left(\frac{V_{a1} - I_{a1} R_a}{V_{a2} - I_{a2} R_a}\right)$$

Speed can be varied by changing the armature voltage, field resistance, or armature resistance.

INDUCTION MOTORS

$$\Omega_s = \text{rotor speed in rad/s} = 2\left(\frac{\omega}{p}\right)$$

ω = electrical frequency in rad/s = $2\pi f$

p = number of poles

stator rms emf per phase = $4.44\, f N_s K_{ws} \phi_p$

ϕ_p = flux per pole

N_s = number of stator conductors per phase

K_{ws} = stator winding factor constant

$0.65 < K_{ws} < 0.86$

The slip, s, is the relative speed between stator rotating field and mechanical speed of rotor.

$$s = \frac{\Omega_s - \Omega_r}{\Omega_s}$$

rotor rms emf per phase = $4.44\, sf N_r K_{wr} \phi_p$

N_r = number of rotor conductors per phase

K_{wr} = rotor winding factor constant

sf = frequency of rotor current

$$\frac{E_r}{E_s} = \frac{sN_r K_{wr}}{N_s K_{ws}}$$

$$E_r = r_r I_r + js\omega L_r I_r$$

The induction motor is modeled like a transformer per phase.

$a = \text{turns ratio} = \dfrac{N_s K_{ws}}{N_r K_{wr}}$

$P_r = \text{power loss per phase in rotor} = I_r^2 r_r$

$R'_m = \text{mechanical load} = r_r \left(\dfrac{1-s}{s}\right)$

$P_g = \text{total gap power} = q I_a^2 r_r \left(\dfrac{1-s}{s}\right)$

$\quad = T\Omega_r = T\Omega_s(1-s)$

$\dfrac{T}{q} = \dfrac{|V_{an}^2|}{\Omega_s}\left[\dfrac{\dfrac{a^2 r_r}{s}}{\left(r_s + \dfrac{a^2 r_r}{s}\right)^2 + X_e^2}\right] = \text{torque per phase}$

q = number of phases

To find the starting torque, let $s = 1$.

$s_{\text{max torque}} = \dfrac{a^2 r_r}{\sqrt{r_s^2 + X_e^2}}$

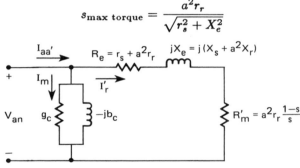

approximate per phase induction motor model with rotor referred to stator

NO-LOAD TEST

This test determines the core losses plus rotational losses.

$g_c = \dfrac{P_{\text{no-load}}}{|V_{an}|^2 q}$

$b_c = \dfrac{Q_{\text{no-load}}}{|V_{an}|^2 q}$

BLOCKED ROTOR TEST

This test determines reactances and resistances.

$R_e = r_s + a^2 r_r = \dfrac{P_{\text{br}}}{I_{an}^2 q}$

$X_e = X_s + a^2 X_r = \dfrac{Q_{\text{br}}}{I_{an}^2 q}$

A third test is necessary to determine r_s and X_s.

REDUCED VOLTAGE STARTING

An autotransformer can be used to supply a reduced voltage for starting. Taps are changed as speed increases.

DELTA-WYE STARTING

The motor is connected wye at starting and is reconnected delta after running. The starting voltage is then 58% of the running voltage across each motor winding.

SYNCHRONOUS MACHINES

A DC current is supplied to the field, which is the rotor through brushes. The armature is a set of stationary windings. The slip is zero and the rotor rotates at armature field frequency.

$\text{number of machine poles} = 120\dfrac{f}{n} = \dfrac{2\omega}{\Omega}$

$\qquad = \dfrac{2 \times \text{electrical frequency}}{\text{mechanical frequency}}$

n = mechanical frequency in rpm
f = electrical frequency in Hz
ω = electrical frequency in rad/s
Ω = mechanical frequency in rad/s

The generated voltage is proportional to the rotor current.

$\text{emf}_{\text{rms per phase}} = E_g = K_\phi I_r$

MOTOR EQUIVALENT CIRCUITS

X_s = synchronous reactance of stator

$E_g = V_a - jX_s I_a$

Let $|E_g| = E$
$\quad |I_a| = I$
$\quad |V_a| = V$

$\dfrac{S}{q} = \dfrac{VE\sin\delta}{X_s} + j\left(\dfrac{V^2 - VE\sin\delta}{X_s}\right)$

$\dfrac{T_{\text{air gap}}}{q} = \dfrac{VE\sin\delta}{X_s \Omega_s}$

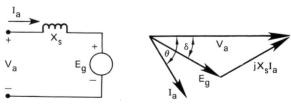

motor equivalent circuit   phasor diagram

θ = power factor angle (+ for current leading voltage)
δ = power angle

ROTATING MACHINES

POWER FACTOR CORRECTION

Adjust rotor current, which controls the value of E_g. For a leading power factor and letting $|E_g| = E$, $|V_a| = V$, $|I_a| = I$,

input power per phase $= \dfrac{P}{q} = VI\cos\theta$

$$E^2 = V^2 + 2X\left(\dfrac{P}{q}\right)\tan\theta + \dfrac{X^2\left(\dfrac{P}{q}\right)^2\sec^2\theta}{V^2}$$

For a lagging power factor, let $\tan\theta$ be negative in the above equation.

A synchronous capacitor is a machine run at no load.

$$\delta = 0°$$

$$\dfrac{Q}{q} = VI = \dfrac{EV - V^2}{X}$$

GENERATOR EQUIVALENT CIRCUIT

Let $|E_g| = E$
$\quad |I_g| = I$
$\quad |V_a| = V$

$$E_g = V_a + jX_s I_a$$

$$\dfrac{S}{q} = \dfrac{EV\sin\delta}{X_s} + j\dfrac{EV\cos\delta - V^2}{X_s}$$

Air gap torque is the same as for a motor.

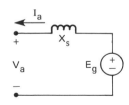

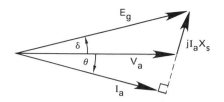

over-excited generator (lagging power factor)

$$|E_g|\cos\delta > V_a$$

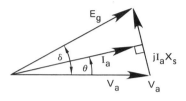

under-excited generator (leading power factor)

$$|E_g|\cos\delta < V_a$$
$$(IX_s)^2 = V^2 + E^2 - 2VE\cos\delta$$
$$E^2 = V^2 + IX_s^2 - 2VIX_s\sin\theta$$

GENERATOR POWER FACTOR

$$\text{power factor} = \cos\theta = \dfrac{I_{a,\min}}{I_a}$$

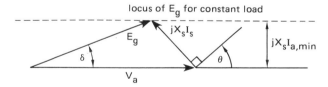

generator vector diagram

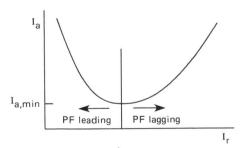

generator vee curve

FUNDAMENTAL SEMICONDUCTOR CIRCUITS

DIODES

A silicon or germanium diode doped with a column III element such as boron is called P-type. Doping with column IV such as phosphorus makes N-type material. When the doping is changed over a small distance, a junction is formed which has the following VA characteristics.

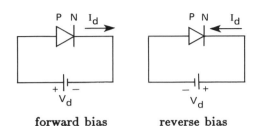

forward bias   reverse bias

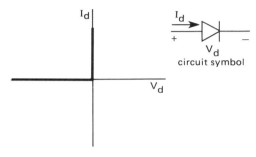

ideal diode

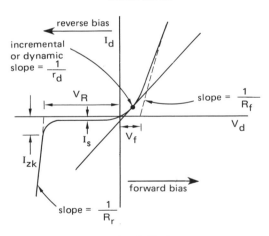

actual diode

V_f = forward conduction (offset) voltage
= 0.2 V for germanium
= 0.6 V for silicon

THEORETICAL DIODE EQUATION

$I_d = I_s \left(e^{V_{PN}/kT} - 1 \right)$

k = Boltzmann constant = 1.38×10^{-23} J/K

T = temperature in K

q = electron charge = 1.602×10^{-19} C

I_s = reverse saturation current

≈ 1% of the forward diode current

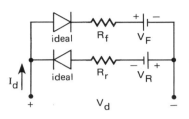

piecewise linear model

PRACTICAL DIODE EQUATION

$I_{PN} = I_S \left(e^{V_{PN}/\eta V_T} - 1 \right)$

V_T = thermal voltage = $(86.21 \times 10^{-6}) T$ V

T = temperature in K

η = a number dependent on the manufacturing process

≈ 2 for silicon, 1 for germanium

INCREMENTAL RESISTANCE

$r_d = \eta \dfrac{V_T}{I_{PN}}$

TEMPERATURE EFFECTS

$V_T = 86.21 \times 10^{-6} T$

I_s doubles every 10°C

$\dfrac{\partial V_{PN}}{\partial T} = -2.5$ mV/°C

ZENER DIODES

A zener diode is one that operates in the reverse bias region of the VA characteristic. A keep-alive current (I_{zk}) required to maintain linear operation. See the diode V characteristic.

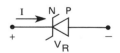

circuit symbol for zener diode

BIPOLAR JUNCTION TRANSISTORS

One of three terminals is common to input and output t form a common base, common emitter, or common collecto circuit.

FUNDAMENTAL SEMICONDUCTOR CIRCUITS

BIASING AND STABILITY

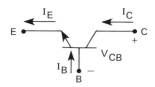

NPN (common base)

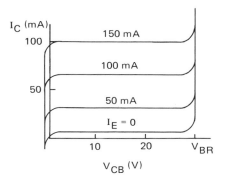

characteristic (common base)

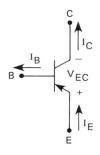

PNP (common emitter)

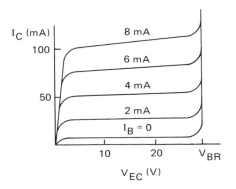

characteristic (common emitter)

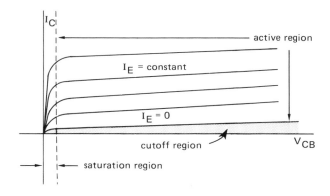

Region	Junction Biasing Emitter/Base	Collector/Base
saturation	forward	forward
active	forward	reverse
cutoff	reverse	reverse

α_o = DC current gain (for a common base)

$$= \frac{dI_C}{dI_E}\bigg|_{V_{CB}} \quad 0.90 \leq \alpha_o \leq 0.99$$

I_{CO} = reverse saturation current of collector with zero emitter current

$I_C = I_{CO} + \alpha_o I_E$

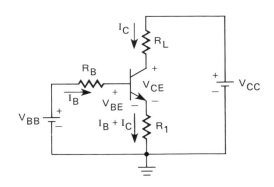

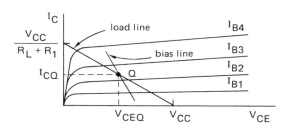

The operating point is established by biasing.

The load line is plotted from the collector loop Kirchhoff's voltage law (for $V_{CE} = 0$, assume $I_B R_1 \ll V_{CC}$).

The collector loop KVL is
$$V_{CC} = V_{CE} + I_B R_1 + I_C(R_1 + R_L)$$
The base loop KVL is
$$V_{BB} - V_{BE} = I_B(R_1 + R_B) + I_C R_1$$
The bias equation is found from the simultaneous solution of the collector and base loop equations, and then plotting the result on the transistor characteristic.

The bias equation is
$$V_{CE} = -\left[\frac{R_1(R_L + R_B) + R_L R_B}{R_1 + R_B}\right] I_C + \left[V_{CC} - \left(\frac{R_1}{R_1 + R_3}\right)(V_{BB} - V_{BE})\right]$$

S = stability factor = $\dfrac{\partial I_C}{\partial I_{CO}}$

S_v = voltage stability factor = $\dfrac{\partial I_C}{\partial T}$

FIXED CURRENT BIAS

$$I_{BQ} = \frac{V_{CC}}{R_B}$$
$$I_C = \frac{I_{CO}}{1 - \alpha_o} + \alpha_o(1 - \alpha_o) I_B$$
$$S = \frac{1}{1 + \alpha_o}$$

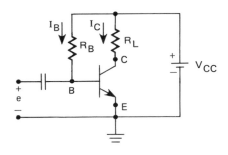

COLLECTOR-TO-BASE BIAS

$$S = \frac{1}{1 - \alpha_o + \dfrac{\alpha_o R_L}{R_L + R_B}}$$

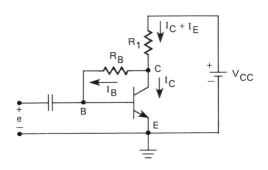

SELF-BIAS

$$S = \frac{1}{1 - \alpha_o + \dfrac{\alpha_o R_1}{R_1 + R_B}} \qquad R_B = \frac{R_2 R_3}{R_2 + R_3}$$

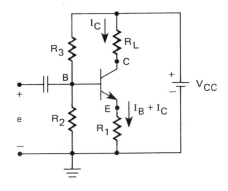

SMALL-SIGNAL TRANSISTOR EQUIVALENT CIRCUITS

Common Base

$$v_{eb} = h_{ib} i_b + h_{rb} v_{cb}$$
$$i_c = h_{fb} i_e + h_{ob} v_{cb}$$

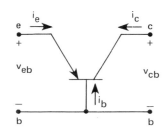

configuration (common base)

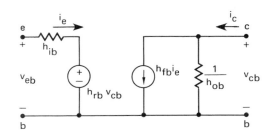

h equivalent circuit (common base)

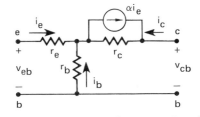

T equivalent circuit (common base)

FUNDAMENTAL SEMICONDUCTOR CIRCUITS

Common Emitters

$v_{be} = h_{ie} i_b + h_{re} v_{ce}$

$i_c = h_{fe} i_b + h_{oe} v_{ce}$

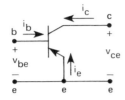

configuration (common emitter)

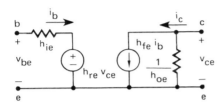

equivalent circuit (common emitter)

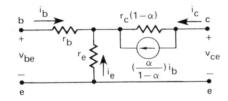

T equivalent circuit (common emitter)

Common Collectors

$v_{bc} = h_{ic} i_b + h_{rc} v_{ec}$

$i_e = h_{fc} i_b + h_{oc} v_{ec}$

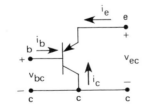

configuration (common collector)

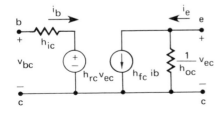

equivalent circuit (common collector)

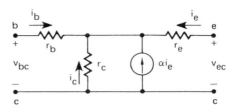

T equivalent circuit (common collector)

PARAMETER CONVERSIONS

To equate parameters, write a set of equations for each circuit, and set coefficients of like independent variables equal. It is faster to use tables from handbooks.

$$\beta = \left.\frac{\partial I_c}{\partial I_b}\right|_{V_{ce}=\text{constant}} = \frac{\alpha_o}{1-\alpha_o} = h_{fe}$$

$$\alpha_o = \frac{h_{fe}}{1+h_{fe}} = \frac{\beta}{\beta+1} = h_{fb}$$

GENERALIZED h PARAMETERS

$V_1 = h_{11} I_1 + h_{12} V_2 = h_i I_1 + h_r V_2$

$I_2 = h_{21} I_1 + h_{22} V_2 = h_f I_1 + h_o V_2$

Note: Another subscript is added to the h parameter when the circuit is common base (b), common emitter (e), or common collector (c).

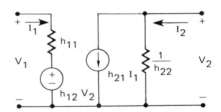

GENERALIZED T EQUIVALENT CIRCUITS

The T circuit can represent a transistor where the collector current is dependent on the emitter current.

$$a = \frac{-\Delta I_2}{\Delta I_1} \text{ with } V_2 = \text{constant}$$

a is similar to α_o.

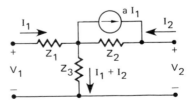

Norton T equivalent circuit

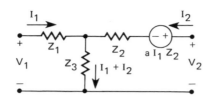

Thevenin T equivalent circuit

MID-FREQUENCY AMPLIFIERS

Impedance of bypass capacitors is assumed to be zero.

COMMON EMITTER AMPLIFIERS

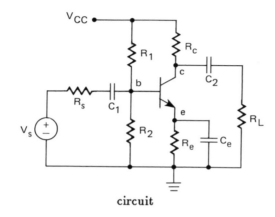

circuit

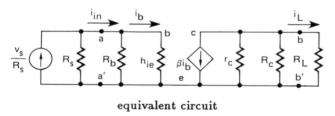

equivalent circuit

The input impedance at terminals a-a' is

$$Z_{in} = h_{ie}\|R_1\|R_2 = h_{ie}\|R_b$$

The output impedance at terminals b-b' is

$$Z_{out} = r_c\|R_c$$

The voltage gain is

$$A_v = \frac{v_{bb'}}{v_{aa'}} = \frac{\beta(r_c\|R_c\|R_L)}{h_{ie}}$$

The current gain is

$$A_i = \frac{i_L}{i_{in}}$$
$$= \frac{-\beta(r_c\|R_c\|R_L)}{R_L} \div \left(1 + \frac{h_{ie}}{R_b}\right)$$
$$R_b = R_1\|R_2$$

COMMON COLLECTOR AMPLIFIERS

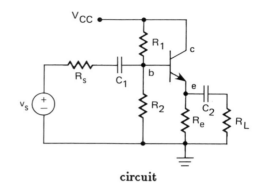

circuit

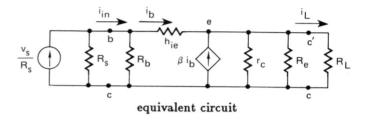

equivalent circuit

The input impedance at terminals b-c is

$$Z_{in} = R_b\|[h_{ie} + (1+\beta)(r_c\|R_e\|R_L)]$$

The output impedance at terminals c'-c is

$$Z_{out} = R_e\|r_c\|\left(\frac{h_{ie} + R_b\|R_s}{1+\beta}\right)$$

The current gain is

$$A_i = \left[\frac{(1+\beta)R_b}{Z_{in}}\right]\left(\frac{r_c\|R_e\|R_L}{R_L}\right)$$

The voltage gain is

$$A_v = \frac{1}{1 + \dfrac{h_{ie}}{(1+\beta)(r_c\|R_e\|R_L)}}$$

COMMON BASE AMPLIFIERS

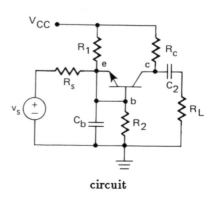

circuit

FUNDAMENTAL SEMICONDUCTOR CIRCUITS

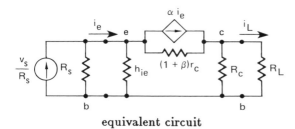

equivalent circuit

The input impedance at terminals e-b is

$$Z_{\text{in}} = \frac{h_{ie}(r_c + R_L\|R_c)}{(1+\beta)r_c + h_{ie} + R_L\|R_c} \approx \frac{h_{ie}}{1+\beta}$$

The output impedance at terminals c-b is

$$Z_{\text{out}} = R_c \| \left[(r_c\|h_{ie}) + R_s + \frac{\beta r_c R_s}{h_{ie} + r_c}\right]$$

$$\approx R_c \| [h_{ie} + (1+\beta)R_s]$$

The current gain is

$$A_i = \left(\frac{R_c}{R_c + R_L}\right)\left[\frac{(1+\beta)r_c + h_{ie}}{(1+\beta)r_c + h_{ie} + R_L\|R_c}\right]$$

$$\approx \frac{\alpha R_c}{R_c + R_L}$$

The voltage gain is

$$A_v = \left(\frac{R_c\|R_L}{h_{ie}}\right)\left(\frac{\beta r_c + h_{ie}}{r_c + h_{ie}}\right) \approx \frac{\beta(R_c\|R_L)}{h_{ie}}$$

JUNCTION FIELD-EFFECT TRANSISTORS

Junction FETs have three layers: a lightly doped substrate, a moderately doped channel, and a heavily doped gate. The gate-channel junction is reverse biased. The arrow on the gate lead points from the P-region to the N-region.

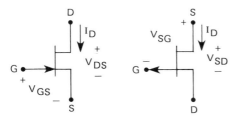

N-channel FET   P-channel FET

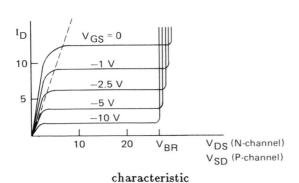

characteristic

METAL-OXIDE SEMICONDUCTOR FIELD-EFFECT TRANSISTORS (MOSFETS)

Depletion mode MOSFETs have the gate insulated from the substrate. Bias voltage can be of either polarity. An enhancement mode MOSFET gate-source voltage is positive for an N-channel and negative for a P-channel, which results in a larger drain current.

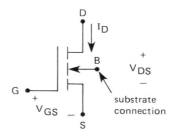

N-channel symbol

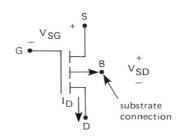

P-channel symbol

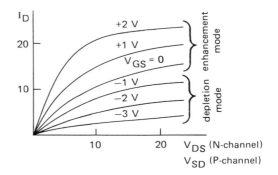

characteristic

SELF-BIASING CIRCUITS

$$R_S = \frac{-V_{\text{GSQ}}}{I_{\text{DQ}}}$$

$$V_{\text{DD}} = V_{\text{DS}} + I_D(R_S + R_D)$$

$$R_D = \frac{V_{\text{DD}} - V_{\text{DSQ}} + V_{\text{GSQ}}}{I_{\text{DQ}}}$$

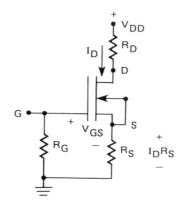

self-bias circuit

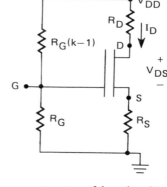

constant g_m bias circuit

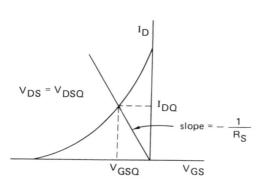

transconductance curve

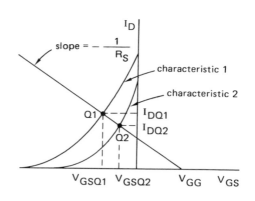

CONSTANT g_m BIAS

It is important to minimize the effect of transconductance variations by using the following equations from two operating points to determine the bias load line.

$$R_D = \frac{V_{DD} - V_{DSQ}}{I_{DQ}} - R_S$$

$$I_{DQ} = 0.5(I_{DQ1} + I_{DQ2})$$

$$k = \frac{V_{DD}}{V_{GG}}$$

$$V_{GS} = V_{GG} - R_S I_D$$

$$V_{GG} = V_{GSQ2} + \left(\frac{V_{GSQ2} - V_{GSQ1}}{I_{DQ1} - I_{DQ2}}\right) I_{DSQ2}$$

$$R_S = \frac{V_{GSQ2} - V_{GSQ1}}{I_{DQ1} - I_{DQ2}}$$

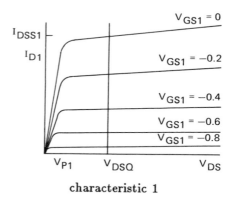

characteristic 1

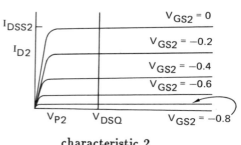

characteristic 2

FUNDAMENTAL SEMICONDUCTOR CIRCUITS

VOLTAGE FEEDBACK BIASING FOR ENHANCEMENT MODE MOSFET

$$V_{DS} = V_{DD} - I_D R_D$$

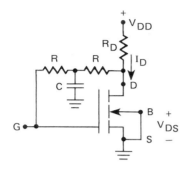

voltage feedback bias circuit

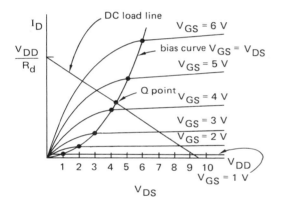

bias operating point on characteristic

CURRENT FEEDBACK BIASING FOR ENHANCEMENT MODE MOSFET

$$V_{GS} = \frac{V_{DD}}{k} - I_D R_S$$

The intersection of the bias curve and the load line is the operating point.

If I_{DQ} is specified, let $k = 2$ and find V_{GSQ} from the transconductance curve and use the following equations.

$$V_{GS} = \frac{V_{DD}}{k} - I_D \left(\frac{\frac{V_{DD}}{k} - V_{GSQ}}{I_{DQ}} \right)$$

$$R_S = \frac{\frac{V_{DD}}{k} - V_{GSQ}}{I_{DQ}}$$

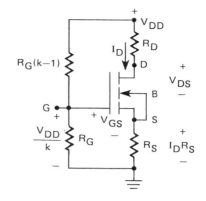

current feedback biasing circuit

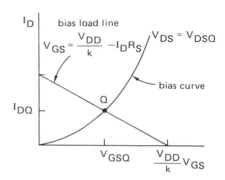

characteristic

COMMON SOURCE FET SMALL-SIGNAL AMPLIFIERS

The gate is electrically isolated from the drain-source circuit.

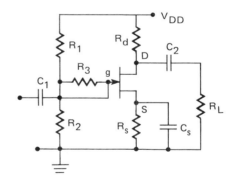

common source FET amplifier circuit

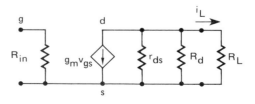

common source small-signal FET equivalent circuit

The input impedance from g to ground is

$$Z_{in} = R_3 + R_1 \| R_2$$

The output impedance from d to ground is

$$Z_{out} = R_d \| r_{ds}$$

The current gain is

$$A_i = -\frac{g_m Z_{in}}{R_L}(r_{ds} \| R_d \| R_L)$$

The voltage gain is

$$A_v = -g_m(r_{ds} \| R_d \| R_L)$$

COMMON DRAIN SMALL-SIGNAL FET AMPLIFIERS

These are also called source followers.

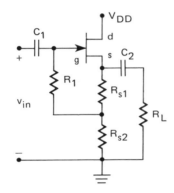

common drain small-signal circuit

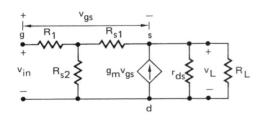

common drain small-signal equivalent circuit

The input impedance from g to ground is

$$Z_{in} = R_1\left(1 + \frac{R_{s2}}{R_{s1}}\right)$$

The output impedance from s to ground is

$$Z_{out} = \frac{1}{g_m + \dfrac{1}{r_{ds}} + \dfrac{1}{R_{s1} + R_{s2}}}$$

The current gain is

$$A_i = g_m R_1 \frac{r_{ds}}{r_{ds} + R_L}\left[\frac{R_{s1} + R_{s2}}{R_{s2} + R(R_L \| r_{ds})(1 + g_m + R_{s1})}\right]$$

The voltage gain is

$$A_v = \frac{1}{1 + \dfrac{(R_{ds} \| R_L) + R_{s1} + R_{s2}}{g_m(r_{ds} \| R_L)(R_{s1} + R_{s2})}}$$

MILLER EQUIVALENT CIRCUIT

The Miller theorem is used to "decouple" an amplifier input from its output so a circuit analysis can be made.

$r_{bb'}$ = base bulk resistance

$C_{b'e}$ = emitter junction capacitance = $\dfrac{1}{\omega_\beta r_e}$

ω_β = frequency at which $\beta = 1$

$C_{b'c}$ = collector junction capacitance

R'_c = the equivalent resistance (including load) across the c-e terminals, including r_c

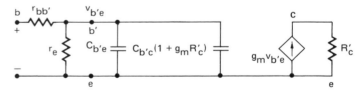

bipolar junction transistor high-frequency circuit

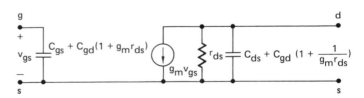

high-frequency FET circuit with Miller capacitance

AMPLIFIER APPLICATIONS

IDEAL OPERATIONAL AMPLIFIERS

An ideal operational amplifier has the following characteristics:

A_o = DC amplifier gain is infinite

Z_{in} = input impedance is infinite

Z_{out} = output impedance is zero

BW = bandwidth is infinite

PRACTICAL OPERATIONAL AMPLIFIERS

$1 \times 10^5 < A_o < 1 \times 10^7$

$R_{in} \simeq 1 \times 10^{12} \, \Omega$

$R_{out} \simeq 1$ to $100 \, \Omega$

$I_{in} \simeq$ nA or pA

For linear operation,

$$|V_+ - V_-| < \frac{V_{DC} - 3}{A_o} = \Delta V$$

Negative feedback is required to prevent instability.

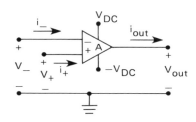

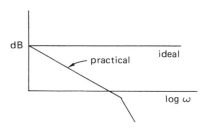

INVERTING OPERATIONAL AMPLIFIERS

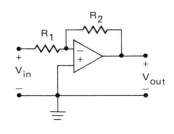

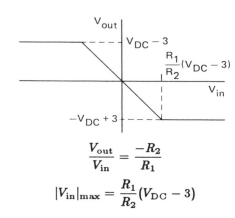

$$\frac{V_{out}}{V_{in}} = \frac{-R_2}{R_1}$$

$$|V_{in}|_{max} = \frac{R_1}{R_2}(V_{DC} - 3)$$

NON-INVERTING OPERATIONAL AMPLIFIERS

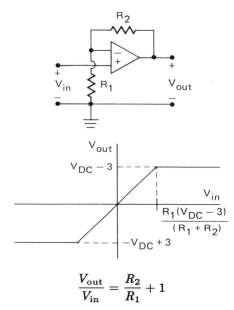

$$\frac{V_{out}}{V_{in}} = \frac{R_2}{R_1} + 1$$

VOLTAGE-FOLLOWER OPERATIONAL AMPLIFIERS

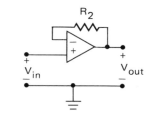

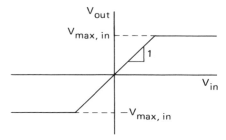

PROFESSIONAL PUBLICATIONS, INC. • Belmont, CA

MULTIPLE-INPUT INVERTER OPERATIONAL AMPLIFIERS (SUMMER)

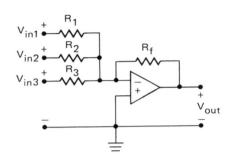

$$V_{\text{out}} = -\left(\frac{R_f}{R_1}\right)V_{\text{in1}} - \left(\frac{R_f}{R_2}\right)V_{\text{in2}} - \left(\frac{R_f}{R_3}\right)V_{\text{in3}}$$

GAIN-BANDWIDTH PRODUCT

GBW = gain-bandwidth product = $A_o \omega_b = \omega_t$
  = (DC gain) (first 3 dB down frequency)
$A_o \simeq 1 \times 10^5$
$\omega_t \simeq 1 \times 10^6$ to 1×10^7 Hz

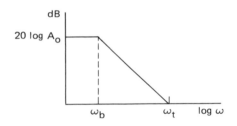

OPERATIONAL AMPLIFIER FREQUENCY RESPONSE (TWO POLES)

A practical operational amplifier has a frequency response as shown on the Bode plot below.

$$\frac{V_{\text{out}}}{\Delta V} = \frac{A_o}{\left(1 + \frac{s}{\omega_b}\right)\left(1 + \frac{s}{\omega_{\text{hf}}}\right)}$$

$$\Delta V = \frac{V_{\text{out}}}{-A_o}\left(1 + \frac{s}{\omega_b}\right)\left(1 + \frac{s}{\omega_{\text{hf}}}\right)$$

$\omega_t = A_o \omega_b$
  = frequency at 0 dB

$$\Delta V = -V_{\text{out}}\left(\frac{1}{A_o} + \frac{s}{\omega_t}\right)\left(1 + \frac{s}{\omega_{\text{hf}}}\right)$$

When $s \ll \omega_{\text{hf}}$,

$$\Delta V = -V_{\text{out}}\left(\frac{1}{A_o} + \frac{s}{\omega_t}\right)$$

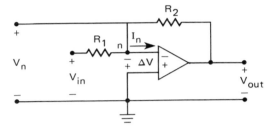

inverting operational amplifier

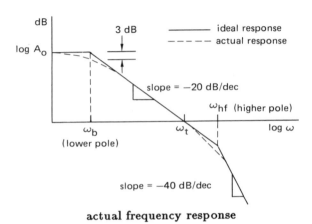

actual frequency response

INVERTING OPERATIONAL AMPLIFIER FREQUENCY RESPONSE

One-Pole, First-Order Approximation

KCL at node n

$$I_n = 0$$

$$\frac{V_{\text{in}} - V_n}{R_1} = \frac{V_n - V_{\text{out}}}{R_2}$$

$$\Delta V = -V_n = V_{\text{out}}\left(\frac{1}{A_o} + \frac{s}{\omega_t}\right)$$

$$\frac{V_{\text{out}}}{V_{\text{in}}} = -\frac{R_2}{R_1}\left[\frac{1}{\left(1 + \frac{s}{\omega_t}\right)\left(1 + \frac{R_2}{R_1}\right)}\right]$$

ω_b = 3 dB corner frequency = $\dfrac{\omega_t}{\left(1 + \dfrac{R_2}{R_1}\right)}$

$$\text{GBW} = A_o \omega_b = \left(\frac{R_2}{R_1 + R_2}\right)\omega_t$$

AMPLIFIER APPLICATIONS

NON-INVERTING OPERATIONAL AMPLIFIER FREQUENCY RESPONSE

One-Pole, First-Order Approximation

KCL at n

$$I_n = 0$$
$$\Delta V = V_{\text{in}} - V_n$$
$$V_n = V_{\text{out}} \left(\frac{1}{A_o} + \frac{s}{\omega_t} \right) = \left(\frac{R_1}{R_1 + R_2} \right) V_{\text{out}}$$

$$\frac{V_{\text{out}}}{V_{\text{in}}} = \frac{1 + \frac{R_2}{R_1}}{1 + \left(1 + \frac{R_2}{R_1}\right) \frac{1}{A_o} + \frac{s}{\omega_t}\left(1 + \frac{R_2}{R_1}\right)}$$

For $A_o \gg 1 + \frac{R_2}{R_1}$,

$$\frac{V_{\text{out}}}{V_{\text{in}}} \simeq \frac{1 + \frac{R_2}{R_1}}{1 + \frac{s}{\omega_t}\left(1 + \frac{R_2}{R_1}\right)}$$

$$\omega_b = \frac{\omega_t}{1 + \frac{R_2}{R_1}}$$

$$\text{GBW} = A_o \omega_b = \omega_t$$

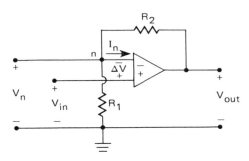

Two-Pole, Second-Order Approximation

$$\frac{V_{\text{out}}}{\Delta V} = -\frac{A_o}{\left(\frac{s}{\omega_b}+1\right)\left(\frac{s}{\omega_{\text{hf}}}+1\right)}$$

$$\frac{V_{\text{out}}}{V_{\text{in}}} = \frac{-\frac{R_2}{R_1}}{1 + \left(1 + \frac{R_2}{R_1}\right)\left[\frac{s^2}{A_o\omega_b\omega_{\text{hf}}} + \frac{s}{A_o}\left(\frac{1}{\omega_b}+\frac{1}{\omega_{\text{hf}}}\right)+\frac{1}{A_o}\right]}$$

For $\omega_t = A_o\omega_b = \text{GBW}$,

$$s = -\frac{\omega_{\text{hf}}+\omega_b}{2} \pm \sqrt{\frac{(\omega_{\text{hf}}-\omega_b)^2}{4} - \frac{\omega_t\omega_{\text{hf}}}{1+\frac{R_2}{R_1}}}$$

For $\omega_{\text{hf}} \simeq \omega_t, \omega_b \ll \omega_t$,

$$s = \frac{\omega_t}{2}\left(-1 \pm \sqrt{1 - \frac{4}{1+\frac{R_2}{R_1}}}\right)$$

When $R_2/R_1 = 3$, $s = -\omega_t/2$ and is real. When $R_2/R_1 > 3$, s is real. When $R_2/R_1 < 3$, the poles are complex conjugates.

ACTIVE FILTERS

Ideal Op-Amp

$$\frac{V_{\text{out}}}{V_{\text{in}}} = -Y_1 Z_2$$
$$\Delta V = 0$$

Practical Op-Amp

A practical op-amp is frequency-limited as follows:

$$\frac{V_{\text{out}}}{V_{\text{in}}} = \frac{-Y_1 Z_2}{1 + \frac{s}{\omega_t}\left(1 + \frac{s}{\omega_{\text{hf}}}\right)(1 + Y_1 Z_2)}$$

$$\Delta V = V_{\text{out}}\left[\frac{s}{\omega_t}\left(1 + \frac{s}{\omega_{\text{hf}}}\right)\right]$$

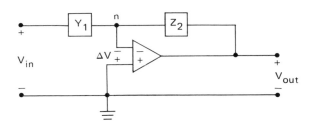

INTEGRATOR OP-AMP FILTERS

Ideal Integrator Op-Amp Filters

$$\frac{V_{\text{out}}}{V_{\text{in}}} = \frac{-1}{sRC}$$

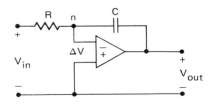

Practical Integrator Op-Amp Filters

A practical op-amp is frequency-limited as follows:

$$\frac{V_{\text{out}}}{V_{\text{in}}} = \frac{\frac{-1}{sRC}}{1 + \frac{s}{\omega_t}\left(1 + \frac{s}{\omega_{\text{hf}}}\right)\left(1 + \frac{1}{sRC}\right)}$$

LOW-PASS OP-AMP FILTERS

Ideal Low-Pass Op-Amp Filters

$$\frac{V_{\text{out}}}{V_{\text{in}}} = \frac{-GR}{sRC + 1}$$

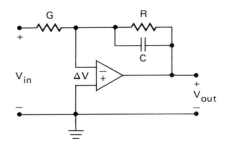

Practical Low-Pass Op-Amp Filters

$$\frac{V_{\text{out}}}{V_{\text{in}}} = \frac{\frac{-GR}{sRC + 1}}{1 + \frac{s}{\omega_t}\left(1 + \frac{s}{\omega_{\text{hf}}}\right)\left(1 + \frac{GR}{sRC + 1}\right)}$$

HIGH-PASS OP-AMP FILTERS

The ideal transfer function is

$$\frac{V_{\text{out}}}{V_{\text{in}}} = \frac{-sC_1 R}{sC_2 R + 1}$$

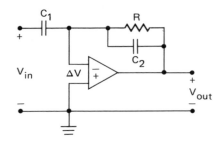

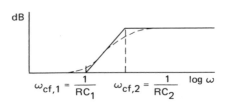

LEAD-LAG OP-AMP FILTERS

$$\frac{V_{\text{out}}}{V_{\text{in}}} = -\left(\frac{R_2}{R_1}\right)\left(\frac{\frac{s}{\omega_z} + 1}{\frac{s}{\omega_p} + 1}\right)$$

$$\omega_z = \frac{1}{R_1 C_1}$$

$$\omega_p = \frac{1}{R_2 C_2}$$

maximum phase shift $= \theta_{\max}$

$$= \arctan\sqrt{\frac{\omega_p}{\omega_z}} - \arctan\sqrt{\frac{\omega_z}{\omega_p}}$$

gain at maximum phase shift $= \dfrac{C_1}{C_2}\sqrt{\dfrac{\omega_z}{\omega_p}}$

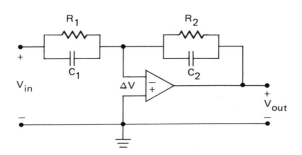

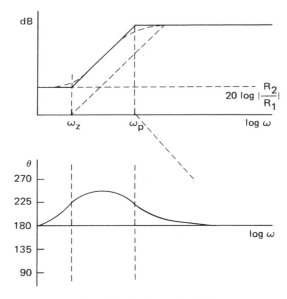

phase lead characteristic

$$\omega_z < \omega_p$$
$$R_1 C_1 > R_2 C_2$$

AMPLIFIER APPLICATIONS

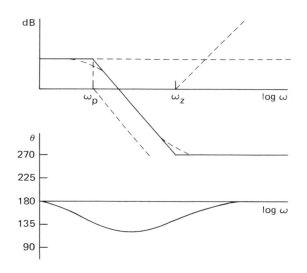

phase lag characteristic

$$\omega_p < \omega_z$$
$$R_2 C_2 > R_1 C_1$$

OP-AMP DIFFERENTIATORS

$$\frac{V_{\text{out}}}{V_{\text{in}}} = \frac{-sC_1 R_2}{(1 + sC_1 R_1)(1 + sC_2 R_2)}$$

For no oscillation to occur, the poles

$$\frac{1}{R_1 C_1} \text{ and } \frac{1}{C_2 R_2} \ll \text{GBW}$$

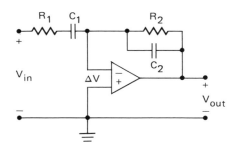

BIQUADRATIC OP-AMP FILTERS

Ideal Op-Amps

$$\Delta V = 0$$

$$\frac{V_{\text{out}}}{V_{\text{in}}} = \frac{-Y_1 Y_3}{Y_5(Y_1 + Y_2 + Y_3 + Y_4) + Y_3 Y_4}$$

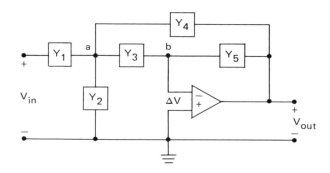

Practical Op-Amps

For $\Delta V = \dfrac{s}{\omega_t} + \dfrac{1}{A_o}\left(\dfrac{s}{\omega_{\text{ht}}} + 1\right)$,

$$\frac{V_{\text{out}}}{V_{\text{in}}} = \frac{-Y_1 Y_3}{D}$$

$$D = \left(1 + \frac{1}{A_o}\right) Y_5 (Y_1 + Y_2 + Y_3 + Y_4) + Y_3 Y_4$$
$$\quad + \frac{Y_3}{A_o}(Y_1 + Y_2 + 2Y_3 + Y_4)$$

BIQUADRATIC LOW-PASS OP-AMP FILTERS

$$\frac{V_{\text{out}}}{V_{\text{in}}} = \frac{-K}{\left(\dfrac{s}{\omega_o}\right)^2 + \dfrac{s}{Q\omega_o} + 1}$$

$$K = G_1 R_4$$

$$\frac{1}{\omega_o^2} = C_2 C_5 R_3 R_4$$

$$\frac{1}{Q\omega_o} = C_5 [R_3(1 + K) + R_4]$$

$$C_2 \geq 4Q^2 (1 + K) C_5$$

$$R_3 = \frac{1}{2Q\omega_o C_5 (1 + K)} \left[1 \pm \sqrt{1 - 4Q^2 \frac{C_5}{C_2}(1 + K)}\right]$$

$$R_4 = \frac{1}{2Q\omega_o C_5} \left[1 \pm \sqrt{1 - 4Q^2 \frac{C_5}{C_2}(1 + K)}\right]$$

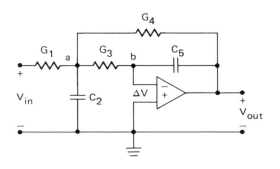

BIQUADRATIC INVERTING TWO-POLE, HIGH-PASS FILTERS

$$\frac{V_{\text{out}}}{V_{\text{in}}} = \frac{C_1 C_3 R_2 R_5 s^2}{s^2 C_3 C_4 R_2 R_5 + s(C_1 + C_3 + C_4) R_2 + 1}$$

$$\frac{1}{\omega_o^2} = C_3 C_4 R_2 R_5$$

$$R_2 = \frac{1}{(C_1 + C_3 + C_4) Q \omega_o}$$

$$R_5 = \frac{Q(C_1 + C_3 + C_4)}{C_3 C_4 \omega_o}$$

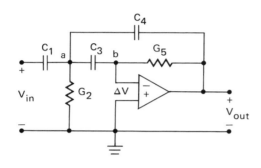

BIQUADRATIC INVERTING BAND-PASS FILTERS (INPUT RESISTOR)

$$\frac{V_{\text{out}}}{V_{\text{in}}} = \frac{-sC_3 R_5}{s^2 C_3 C_4 R_1 R_5 + s(C_3 + C_4) R_1 + 1}$$

$$s_{1,2} = (0.5)\left(\frac{C_3 + C_4}{C_3 C_4 C_5}\right)\left[1 \pm \sqrt{1 - \left(\frac{4 C_3 C_4}{(C_3 + C_4)^2}\right)\left(\frac{R_5}{R_1}\right)}\right]$$

Letting $C_3 = C_4 = C$, the corner frequencies are

$$s_{1,2} = -\frac{1}{CR_5}\left(1 \pm \sqrt{1 - \frac{R_5}{R_1}}\right)$$

The roots can be real or complex.

$$\omega_{\text{mid-band}} = \sqrt{\omega_1 \omega_2}$$

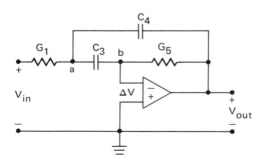

BIQUADRATIC INVERTING NOTCH FILTERS WITH Y_1 RESISTIVE

$$\frac{V_{\text{out}}}{V_{\text{in}}} = \frac{s^2 C_3 C_4 R_1 R_5 + 1}{k[s^2 C_3 C_4 R_1 R_5 + R_1 s(C_3 + C_4) + 1]}$$

$\omega_n = $ notch frequency

$$\omega_n^2 = \frac{1}{C_3 C_4 R_1 R_5}$$

$$Q = \sqrt{\frac{R_5}{R_1}} \frac{\sqrt{C_3 C_4}}{C_3 + C_4}$$

$$\text{bandwidth} = \text{BW} = \omega_2 - \omega_1 = \frac{\omega_n}{Q}\sqrt{1 + 4Q^2}$$

Letting $C_3 = C_4$,

$$\frac{\omega_2 - \omega_1}{\omega_n} = \sqrt{1 + \frac{R_5}{R_1}}$$

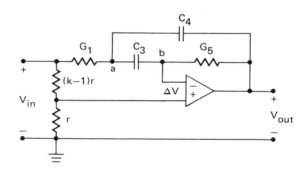

SALLEN AND KEY LOW-PASS FILTERS

$$\frac{V_{\text{out}}}{V_{\text{in}}} = $$

$$\frac{k}{s^2 C_4 C_5 R_1 R_3 + s[C_5(R_1 + R_3) + (1 - k)C_4 R_1] + 1}$$

$$\omega^2 = \frac{1}{C_4 C_5 R_1 R_3}$$

$$\frac{1}{Q} = \sqrt{\frac{C_5}{C_4}}\left(\frac{R_1 + R_5}{\sqrt{R_1 R_3}}\right) - (k-1)\sqrt{\left(\frac{C_4}{C_5}\right)\left(\frac{R_1}{R_3}\right)}$$

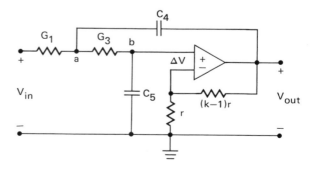

AMPLIFIER APPLICATIONS

SALLEN AND KEY HIGH-PASS FILTERS

$$\frac{V_{\text{out}}}{V_{\text{in}}} =$$

$$\frac{s^2 k C_1 C_3 R_4 R_5}{s^2 C_1 C_3 R_4 R_5 + s[(C_1 + C_3)R_4 + (1-k)C_3 R_5] + 1}$$

$$\omega_o^2 = \frac{1}{C_1 C_3 R_4 R_5}$$

$$\frac{1}{Q} = \frac{\sqrt{\frac{R_4}{R_5}}}{\sqrt{C_1 C_3}}(C_1 + C_3) - (k-1)\sqrt{\left(\frac{C_3}{C_1}\right)\left(\frac{R_5}{R_4}\right)}$$

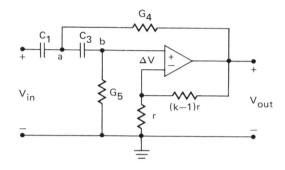

SALLEN AND KEY BAND-PASS FILTERS

$$\frac{V_{\text{out}}}{V_{\text{in}}} =$$

$$\frac{skC_3 R_4 R_5}{s^2 C_2 C_3 R_1 R_4 R_5 + s(C_2 + C_3)R_1 R_4 + C_3 R_5[R_4 + (1-k)R_1] + R_1 + R_4}$$

$$\omega_o^2 = \frac{R_1 + R_4}{C_2 C_3 R_1 R_4 R_5}$$

$$\frac{1}{Q} = \frac{C_2 + C_3}{\sqrt{C_2 C_3}}\sqrt{\frac{R_1 R_4}{R_5(R_1 + R_4)}} + \sqrt{\frac{C_3 R_5}{C_2 R_4}}\sqrt{\frac{R_1}{R_1 + R_4}}$$
$$\times \left(\frac{R_4}{R_1} + 1 - k\right)$$

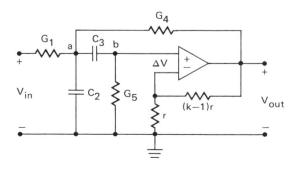

OPERATIONAL AMPLIFIER INTERNAL CHARACTERISTICS

$$\frac{V_o}{V_1} = \frac{\mu g_m r}{sCr(1+\mu) + 1}$$

$$A_o = \mu g_m r$$

$$\omega_b = \frac{1}{Cr(1+\mu)}$$

$$\omega_t = A_o \omega_b \quad \mu \gg 1$$

$$\omega_t = \frac{\mu g_m r}{Cr(1+\mu)} \simeq \frac{g_m}{C}$$

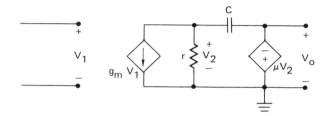

For large signal input (saturating amplifier),

$$V_o(t) \simeq \mu r I t \omega_b$$

$$I = \text{average current}$$

$$\text{slew rate} = \frac{I}{C} = \frac{dV_o(t)}{dt} = S_R$$

= limit of slope on output voltage

$$\omega_{\text{max}} = \frac{S_R}{V_{\text{max}}}$$

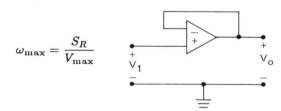

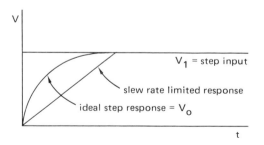

$$V_o(t) = V_s(1 - e^{-\omega_t t})$$

$$V_{\text{max}} \text{ sinusoidal} = V_{\text{DC}} - 3 \text{ V}$$

COMMON MODE GAIN

An ideal op-amp does not amplify a signal that is identical at each input (common mode). A practical op-amp has a common-mode amplification.

CMRR = common-mode rejection ratio in dB

$$= \frac{\text{differential signal gain}}{\text{common signal gain}}$$

WAVESHAPING, LOGIC, AND DATA CONVERSION

PRECISION DIODES

The precision diode eliminates the "offset voltage" by using a high-gain operational amplifier.

$$V_L = \frac{R_o V_{\text{in}}}{R_o + \frac{R_o + R_f}{A_v}} - \frac{R_o \frac{V_f}{A_v}}{R_o + \frac{R_o + R_f}{A_v}}$$

When $A_v \simeq 1 \times 10^5$,

$$V_L \simeq V_{\text{in}} - \frac{V_f}{A_v}$$

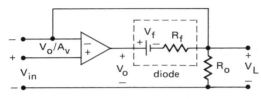

precision diode circuit

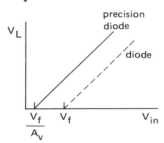

voltage characteristic

CLIPPER CIRCUITS

The clipper circuit "clips" the voltage to a reference value. V_r can be positive or negative.

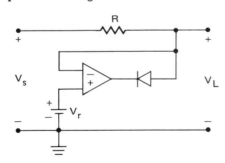

precision clipper circuit

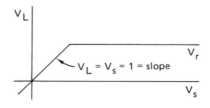

voltage characteristic

LIMITER CIRCUITS

The limiter circuit "limits" the voltage to some reference value. V_r can be positive or negative.

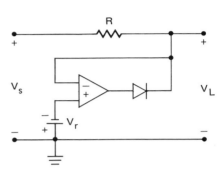

precision limiter

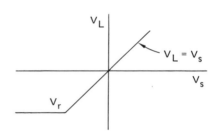

voltage characteristic

ZENER REGULATORS

A zener diode voltage regulator is constructed of a series resistor and shunt zener diode. Voltage regulation is accomplished by the zener diode conducting excess load current.

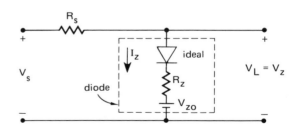

zener diode equivalent circuit

I_{zk} = keep-alive current

$$I_{zk} = \frac{V_L - V_{zo}}{R_z}$$

V_{zo} = forward bias conduction voltage

$$R_s \leq \frac{V_{s,\text{min}} - V_{zo}}{I_{L,\text{max}}}$$

$$V_L = \frac{V_{zo} R_s + V_s R_z}{R_s + R_z} - I_L \left(\frac{R_s R_z}{R_s + R_z} \right)$$

WAVESHAPING, LOGIC, AND DATA CONVERSION

OPERATIONAL AMPLIFIER REGULATORS

Output voltage regulation can be improved by using an operational amplifier in conjunction with a pass transistor.

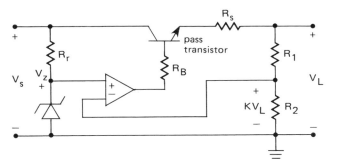

operational amplifier voltage regulator circuit

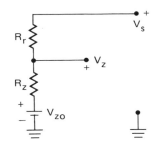

reference voltage circuit

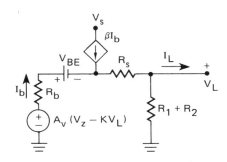

output equivalent circuit

$$\% \text{ regulation} = \frac{V_{z,\max} - V_{z,\min}}{V_{z,\min}} \times 100\%$$

$$V_L = \frac{A_v V_z - V_{BE} - I_L \left(R_s + \frac{R_b}{\beta+1}\right)}{KA_v + 1 + \frac{K}{R_2}\left(R_s + \frac{R_b}{\beta+1}\right)}$$

$$\simeq \frac{V_z}{K} - I_L \frac{R_s + \frac{R_b}{\beta+1}}{KA_v}$$

set $V_{CE} = 1$ V

$$K = \frac{R_2}{R_1 + R_2}$$

$$R_r = \frac{V_{s,\min} - V_{zo}}{I_{zk}} - R_z$$

LOGIC CIRCUITS

Transistor logic circuits toggle between the volt-ampere characteristic saturation region and cutoff region.

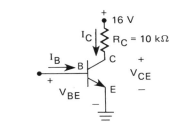

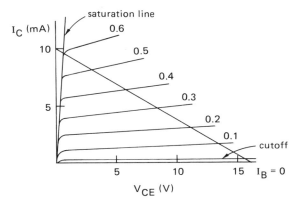

TRANSISTOR LOGICAL NOT INVERTERS

The common-emitter transistor circuit is a NOT inverter as shown by the truth table.

Input V_{BE}	Output V_{CE}	Operating Region
low	high	cutoff
high	low	saturation

RESISTOR-TRANSISTOR (RTL) NOR GATES

H = high voltage $\simeq 3$ V
L = low voltage $\simeq 0.2$ V (forward base to emitter junction conduction voltage)

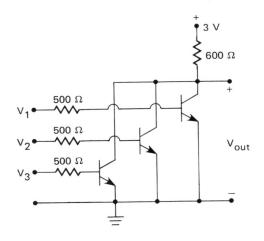

V_1	V_2	V_3	V_{out}
L	L	L	H
L	L	H	L
L	H	L	L
L	H	H	L
H	L	L	L
H	L	H	L
H	H	L	L
H	H	H	L

DIODE-TRANSISTOR (DTL) AND GATES

Two diodes are connected in series with the base to assure the transistor does not turn on with one input low.

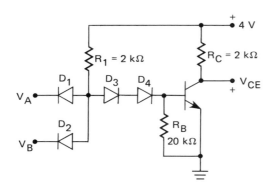

V_A	V_B	V_{CE}
L	L	H
L	H	H
H	L	H
H	H	L

TRANSISTOR-TRANSISTOR NAND LOGIC GATES

A multi-emitter transistor drives the base of T2. The three junctions require 1.8 V for forward biasing. When both inputs are high, the base of T1 connects to V_{CC}. This causes forward biasing of T1, T2, and T3 junctions to ground.

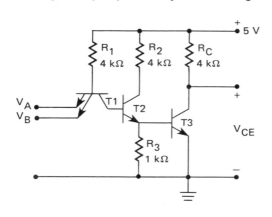

V_A	V_B	V_{CE}
L	L	H
L	H	H
H	L	H
H	H	L

FAN-IN/FAN-OUT

A gate's fan-in rating is the total sink current or maximum number of gates allowed to be connected to the input.

A gate fan-out rating is the total source current or maximum number of gates allowed to be connected to the output.

The total current provided by the source device is called the source current. The total current provided by the sunk device is called the sink current.

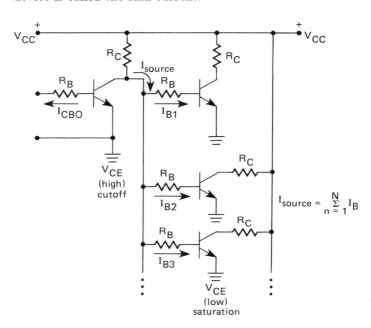

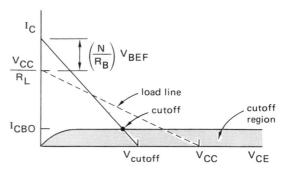

operating point for sourcing N transistors

I_{CBO} = collector-base reverse-bias current (cutoff)

$V_{BEF} \simeq 0.6$ V

Source transistor is at cutoff (high collector voltage). Driven transistors are at saturation (low collector voltage).

$$N \leq \frac{V_{CC} - V_{BEF} - R_C I_{CBO}}{I_{B,\min} R_C} - \frac{R_B}{R_C}$$

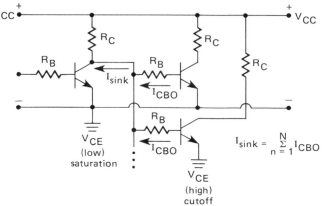

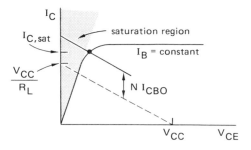

operating point for sinking N transistors

$$N \leq \frac{R_C I_{C,\text{sat}} + V_{CE,\text{sat}} - V_{CC}}{R_C I_{CBO}}$$

GATE TIME DELAYS

t_d = delay time $\simeq$ 1 to 10 ns (typical 3 ns)
t_r = rise time $\simeq$ 1 to 100 ns (typical 50 ns)
t_s = storage time $\simeq$ 5 to 50 ns (typical 25 ns)
t_f = fall time $\simeq$ 1 to 100 ns (typical 30 ns)

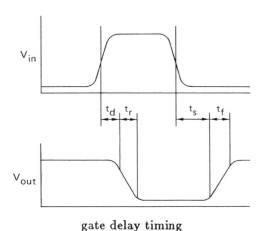

gate delay timing

POWER SWITCHING

UJT-SCR Trigger Circuits

The symbol and VA characteristic of a unijunction transistor (UJT), also called a double-base diode, is shown here.

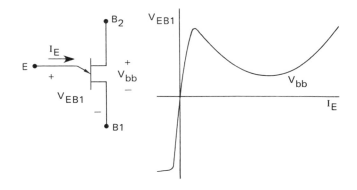

UJT circuit symbol   UJT characteristic

A silicon-controlled rectifier is a form of diode that has a gate that can turn on the diode. A DC current must be applied for about 50 μs.

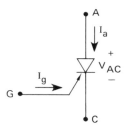

SCR circuit symbol

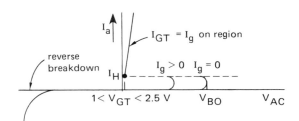

SCR characteristic

A UJT can be used to provide SCR gate current in the following circuit. $R_1 + R_2$ (base resistances) varies between 1000 to 10,000 Ω; a typical value is 5000 Ω.

I_H = minimum current to keep UJT triggered or "on" condition

$$V_E(\text{trigger}) = V_{BB} \frac{R_1 + R_{B1}}{R_1 + R_{B1} + R_2} + 0.6$$

$$Q = \text{charge} = I_{GT}(50 \times 10^{-6}) = C_E V_{\text{initial}} - V_{GT}$$

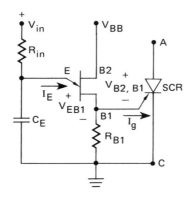

UJT-SCR trigger circuit

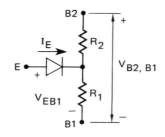

UJT pre-trigger circuit

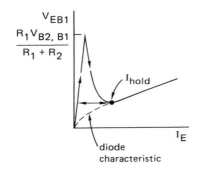

DIAC TRIGGER FOR TRIAC

A DIAC is a back-to-back diode with no trigger gate.

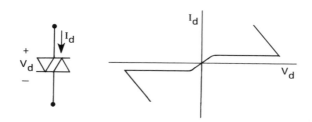

DIAC symbol   DIAC characteristic

A TRIAC is two back-to-back SCRs with trigger gates connected together.

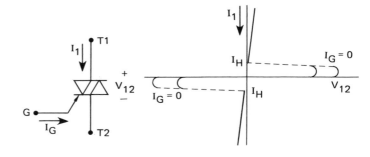

TRIAC symbol   TRIAC characteristic

A DIAC-triggered TRIAC circuit is shown. R controls the firing angle and C releases charge to swing the DIAC from the breakover voltage to the dynamic on-resistance.

$$V_{C,\text{ss}} = \frac{V_s}{jwRC+1}$$

Use iteration to solve for $V_{C,\text{ss}}$ from the following equations.

$$V_{C,\text{ss}} = \frac{V_{s,\max}}{\sqrt{\omega^2 R^2 C^2 + 1}} \sin(\omega t_f - \arctan \omega RC) = V_{BO}$$

$$V_{C,\text{ss}} = V_{C,\text{ss}} + \left[V_C(0) + \frac{V_{s,\max} \omega RC}{\omega^2 R^2 C^2 + 1}\right] \exp^{-t/RC}$$

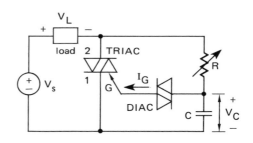

DIAC-triggered TRIAC circuit

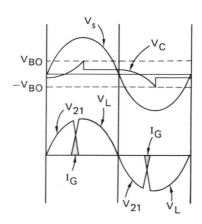

DIAC trigger waveform

WAVESHAPING, LOGIC, AND DATA CONVERSION

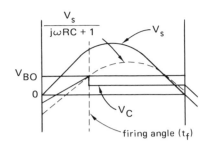

capacitor waveform

UJT RELAXATION OSCILLATORS

$$V_C = V_{bb}(1 - \exp^{-t/\tau}) = V_E$$
$$\tau = RC$$

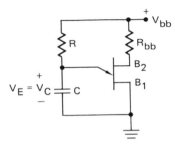

UJT relaxation circuit

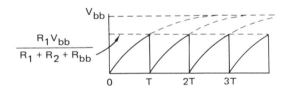

waveform across C

NEON BULB RELAXATION OSCILLATORS

At a specified voltage the neon bulb ionizes, causing conduction.

$$V_o = \frac{V_C R_1}{R_1 + R_2}$$
$$V_{Th} = V_s \left(\frac{R_1 + R_2}{R_1 + R_2 + R_s} \right)$$
$$R_{Th} = R_s \left(\frac{R_1 + R_2}{R_1 + R_2 + R_s} \right)$$

Prior to ionization,

$$V_C = V_{Th}(1 - \exp^{-t/\tau})$$
$$\tau = R_{Th} C$$
$$T = R_{Th} C \ln \left(\frac{V_{Th}}{V_{Th} - V_I} \right)$$

V_I = neon ionization voltage

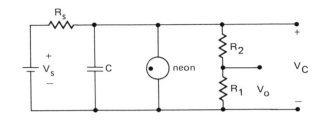

neon bulb relaxation oscillator

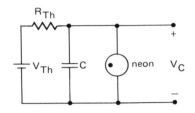

Thevenin circuit

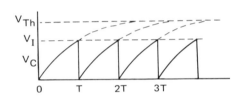

capacitance waveform

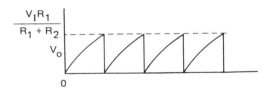

output waveform

OPERATIONAL AMPLIFIER SQUARE-WAVE GENERATORS

$$k > 1$$
$$T_{s1} = R_1 C \ln \left[\frac{kV_L + V_H}{V_L(k-1)} \right]$$
$$T_{s2} = R_2 C \ln \left[\frac{kV_H + V_L}{V_H(k-1)} \right]$$

To obtain a symmetrical wave, let $R_1 = R_2$ and $V_H = V_L$.

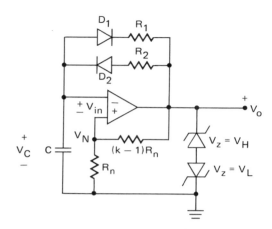

square-wave generator

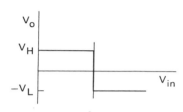

op-amp characteristic

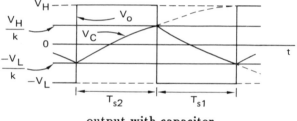

output with capacitor

ASTABLE MULTIVIBRATORS

An astable multivibrator is an amplifier circuit connected so that it oscillates between two stable states. The collector-coupled transistor circuit shown oscillates by the repeated charging and discharging of the collector capacitors.

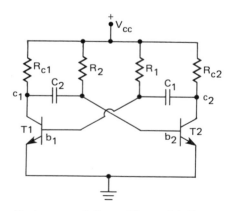

collector-coupled astable multivibrator

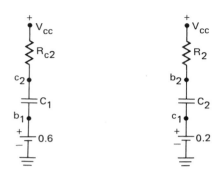

C_1 charging circuit C_2 charging circuit

Letting $R_2 = R_{c2}$, the equations can be solved algebraically. Otherwise an iteration method must be used.

$$\frac{T}{2R_{c2}C} = \ln\left(\frac{V_{cc} - 0.2}{V_{cc} - 0.6 - V_{C2}(0)}\right)$$

$$\frac{T}{2R_2C} = \ln\left(\frac{V_{cc} - 0.2 + V_{C2}(0)}{V_{cc} - 0.6}\right)$$

MONOSTABLE MULTIVIBRATORS

A monostable multivibrator has one stable state. It must be triggered.

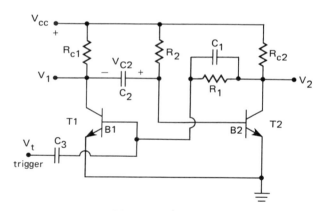

monostable multivibrator circuit

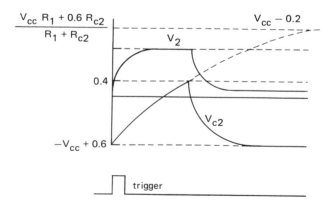

monostable waveforms

DIGITAL LOGIC

BOOLEAN ALGEBRA

A logical variable can have one of two possible states known as true (1) or false (0). A logical constant can have the value of 1 or 0.

Function Binary Decimal	00 0	XY Variable Combination 01 10 11 1 2 3			Logic Gate	Logic Operation
F_0	0	0	0	0		
F_1	0	0	0	1	AND	$X \cdot Y$
F_2	0	0	1	0		
F_3	0	0	1	1		
F_4	0	1	0	0		
F_5	0	1	0	1		
F_6	0	1	1	0	XOR	$X \oplus Y$
F_7	0	1	1	1	OR	$X + Y$
F_8	1	0	0	0	NOR	$\overline{X + Y}$
F_9	1	0	0	1	XNOR	$\overline{X \oplus Y}$
F_{10}	1	0	1	0	NOT Y	$\overline{Y}$
F_{11}	1	0	1	1		
F_{12}	1	1	0	0	NOT X	$\overline{X}$
F_{13}	1	1	0	1		
F_{14}	1	1	1	0	NAND	$\overline{X \cdot Y}$
F_{15}	1	1	1	1		

FUNCTIONS OF TWO VARIABLES

In general, there are 2^n possible combinations of n variables and 2^{2^n} functions.

LOGICAL IDENTITIES

Relationships Between Constants

$$0 \cdot 0 = 0$$
$$0 \cdot 1 = 0$$
$$1 \cdot 1 = 1$$
$$0 + 0 = 0$$
$$0 + 1 = 1$$
$$1 + 1 = 1$$
$$\overline{0} = 1$$
$$\overline{1} = 0$$
$$0 \oplus 0 = 0$$
$$0 \oplus 1 = 1$$
$$1 \oplus 0 = 1$$
$$1 \oplus 1 = 0$$

Relationships Between Variables

$$A + B = B + A$$
$$A \cdot B = B \cdot A$$
$$A + \overline{A} = 1$$
$$A \cdot \overline{A} = 0$$
$$A + A = A$$
$$A \cdot A = A$$
$$\overline{\overline{A}} = A$$
$$A + (B + C) = A + B + C$$
$$A \cdot (B + C) = (A \cdot B) + (A \cdot C)$$
$$A + (B \cdot C) = A + B \cdot C$$
$$A \oplus B = \overline{A} \cdot B + A \cdot \overline{B}$$
$$A \odot B = \overline{A} \cdot \overline{B} + A \cdot B$$
$$A \oplus B \oplus C = A \oplus (B \oplus C)$$
$$= (A \oplus B) \oplus C$$

LOGICAL THEOREMS

$$A + A \cdot B = A$$
$$A + \overline{A} \cdot B = A + B$$
$$A \cdot C + \overline{A} \cdot B \cdot C = A \cdot C + B \cdot C$$
$$A \cdot B + A \cdot C + \overline{B} \cdot C = A \cdot B + \overline{B} \cdot C$$
$$A \cdot B + A \cdot \overline{B} = A \text{ (logical adjacency)}$$
$$\overline{A \cdot B \cdot C \cdot D \cdot E} = \overline{A} + \overline{B} + \overline{C} + \overline{D} + \overline{E} \text{ (de Morgan)}$$
$$\overline{A + B + C + D + E} = \overline{A} \cdot \overline{B} \cdot \overline{C} \cdot \overline{D} \cdot \overline{E} \text{ (de Morgan)}$$

LOGIC GATES AND TRUTH TABLES

AND Gate

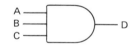

$$D = A \cdot B \cdot C$$

Truth Tables for AND Gate

A	B	C	D
0	0	0	0
0	0	1	0
0	1	0	0
0	1	1	0
1	0	0	0
1	0	1	0
1	1	0	0
1	1	1	1

NAND Gate

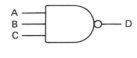

$$D = \overline{A \cdot B \cdot C}$$

Truth Table for NAND Gate

A	B	C	D
0	0	0	1
0	0	1	1
0	1	0	1
0	1	1	1
1	0	0	1
1	0	1	1
1	1	0	1
1	1	1	0

NOR Gate

$$D = \overline{A + B + C}$$

Truth Table for NOR Gate

A	B	C	D
0	0	0	1
0	0	1	0
0	1	0	0
0	1	1	0
1	0	0	0
1	0	1	0
1	1	0	0
1	1	1	0

NOT Gate

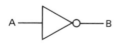

$$B = \overline{A}$$

Truth Table for NOT Gate

A	B
0	1
1	0

OR Gate

$$D = A + B + C$$

Truth Table for OR Gate

A	B	C	D
0	0	0	0
0	0	1	1
0	1	0	1
0	1	1	1
1	0	0	1
1	0	1	1
1	1	0	1
1	1	1	1

XOR Gate

$$C = A \oplus B$$

Truth Table for XOR Gate

A	B	C
0	0	0
0	1	1
1	0	1
1	1	0

XNOR Gate

$$C = \overline{A \oplus B} = A \odot B$$

Truth Table for XNOR Gate

A	B	C
0	0	1
0	1	0
1	0	0
1	1	1

Alternate XNOR (coincidence gate symbol)

MINTERMS AND MAXTERMS

Variable Combinations $A\ B$	Product Terms (Minterms)	Sum Term (Maxterms)
0 0	$\overline{A} \cdot \overline{B} = m_{00}$	$A + B = M_{00}$
0 1	$\overline{A} \cdot B = m_{01}$	$A + \overline{B} = M_{01}$
1 0	$A \cdot \overline{B} = m_{10}$	$\overline{A} + B = M_{10}$
1 1	$A \cdot B = m_{11}$	$\overline{A} + \overline{B} = M_{11}$

Minterms are variables "ANDed." Where $C = AB$ is true (1), the minterms are "ORed" to obtain the sum of products (SOP) form.

$$C = m_{00} + m_{01} + m_{10} + m_{11}$$
$$C = (\overline{A} \cdot \overline{B}) + (\overline{A} \cdot B) + (A \cdot \overline{B}) + (A \cdot B)$$

Maxterms are "NOTed" variables "ORed." Where C is false (0), the maxterms are "ANDed" to obtain the product of sum (POS) form.

$$C = M_{00} + M_{01} + M_{10} + M_{11}$$
$$C = (A + B)(A + \overline{B})(\overline{A} + B)(\overline{A} + \overline{B})$$

CANONICAL REALIZATION

When all variables or their complements appear only once in each term of a logical expression, that expression is said to be in canonical form. A realization is the design of a logical gate circuit from the canonical form. The following steps illustrate the process.

step 1: For a given truth table, complete the maxterms and minterms. Minterms are where $F = 1$ (true). Maxterms are where $F = 0$ (false).

XYZ	F_4	m	M
0 0 0	0	–	$X + Y + Z$
0 0 1	1	$\overline{X} \cdot \overline{Y} \cdot Z$	–
0 1 0	1	$\overline{X} \cdot Y \cdot \overline{Z}$	–
0 1 1	0	–	$X + \overline{Y} + \overline{Z}$
1 0 0	1	$X \cdot \overline{Y} \cdot \overline{Z}$	–
1 0 1	0	–	$\overline{X} + Y + \overline{Z}$
1 1 0	0	–	$\overline{X} + \overline{Y} + Z$
1 1 1	1	$X \cdot Y \cdot Z$	–

step 2: Combine the minterms and maxterms to get the SOP and POS expressions.

SOP $F_4 = \overline{X}\,\overline{Y}Z + \overline{X}Y\overline{Z} + X\overline{Y}\,\overline{Z} + XYZ$
POS $F_4 = (X + Y + Z)(X + \overline{Y} + \overline{Z})$
$(\overline{X} + Y + \overline{Z})(\overline{X} + \overline{Y} + Z)$

step 3: Design the circuit using AND and OR gates working from the inside of the SOP or POS expressions. The use of other gates is also possible.

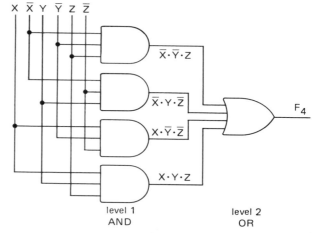

SOP AND/OR realization

MINIMIZATION USING KARNAUGH MAPS

To minimize a function, follow these steps:

step 1: Construct a Karnaugh map from the truth table.

Truth Table

cell number	A	B	C	D
1	0	0	0	0
2	0	0	1	0
3	0	1	0	1
4	0	1	1	0
5	1	0	0	1
6	1	0	1	0
7	1	1	0	1
8	1	1	1	0

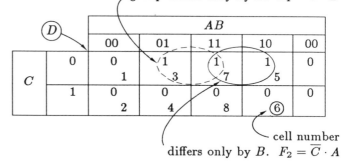

Karnaugh Map

step 2: Combine adjacent terms into the largest possible groupings. SOP are 1's and are easiest to work with.

step 3: Include a sufficient number of groupings to include each of the terms in at least one group.

step 4: "OR" the SOP terms together to obtain the logical expression.

$$D = \overline{C}B + \overline{C}A$$

step 5: Check the expression against the truth table.

DATA SELECTORS

Data selectors are also called multiplexers or MUXs.

input		functions		
A	B	$\overline{A} \cdot B$	$A + \overline{B}$	$A \oplus B$
0	0	0	1	0
0	1	1	0	1
1	0	0	1	1
1	1	0	1	0

select input	residues		
	$\overline{A} \cdot B$	$A + \overline{B}$	$A \oplus B$
$\overline{A}$	B	$\overline{B}$	B
A	0	1	$\overline{B}$

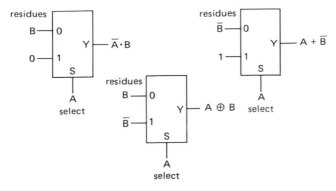

2-input MUX logic

AB	C	Y	residues
0 0	0	0	
0 0	1	0	0
0 1	0	1	
0 1	1	0	$\overline{C}$
1 0	0	0	
1 0	1	1	C
1 1	0	1	
1 1	1	1	1

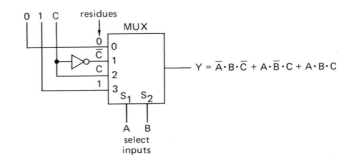

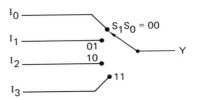

4-input data selector realization

FLIP-FLOPS

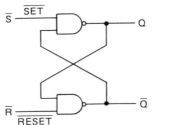

NAND S-R flip-flop

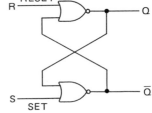

NOR S-R flip-flop

Next-State Table for S-R Flip-Flop

present			next
S	R	Q	Q
0	0	0	0
0	1	0	0
1	0	0	1
1	1	0	N
0	0	1	1
0	1	1	0
1	0	1	1
1	1	1	N

N = input not permitted

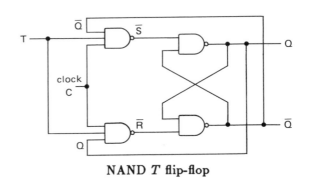

NAND T flip-flop

DIGITAL LOGIC

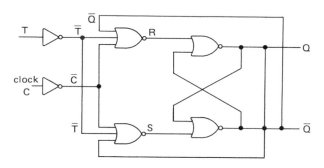

NOR T flip-flop

Transition Table for T Flip-Flop

present Q	next Q	T	$\overline{S}$	$\overline{R}$
0	0	0	1	X
0	1	1	0	1
1	0	1	1	0
1	1	0	X	1

X = don't care

$S = \overline{Q} \cdot T$

$S = \overline{\overline{Q} \cdot T}$

Transition Table for D Flip-Flop

E	D	present Q	next Q	S	R
1	0	0	0	0	X
1	0	1	0	0	1
1	1	0	1	1	0
1	1	1	1	X	0
0	X	X	present	0	0

E = enable input

Truth Table for J-K Flip-Flop

J	K	present Q	next Q	S	R
0	0	0	0	0	X
0	0	1	1	X	0
0	1	0	0	0	X
0	1	1	0	0	1
1	0	0	1	1	0
1	0	1	1	X	0
1	1	0	1	1	0
1	1	1	0	0	1

$S = J \cdot \overline{Q}$

$R = K \cdot Q$

Transition Table for Flip-Flop

present Q	next Q^+	S	R	J	K	T	D
0	$\longrightarrow$ 0	0	X	0	X	0	0
0	$\longrightarrow$ 1	1	0	1	X	1	1
1	$\longrightarrow$ 0	0	1	X	1	1	0
1	$\longrightarrow$ 1	X	0	X	0	0	1

$$Q^+ = S + \overline{R} \cdot Q$$
$$= J \cdot \overline{Q} + \overline{K}$$
$$= T \oplus Q$$
$$= D$$

COUNTERS

Counters are sets of flip-flops. The count is 2^n where n = number of flip-flops in the circuit.

A synchronous counter has each flip-flop state changed at the same time by a clock.

Asynchronous counters do not have simultaneous clocked inputs. An example is the ripple counter.

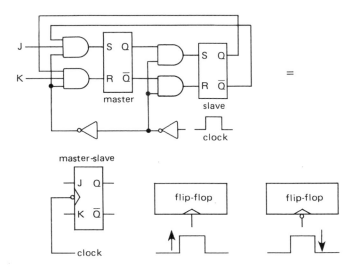

master-slave J-K flip-flop

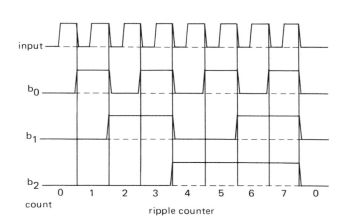

timing diagram

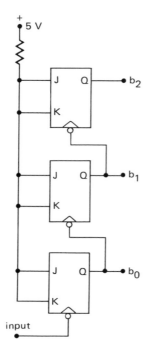

ripple counter circuit

COUNTER ANALYSIS

step 1: Write the logical expressions from the circuit.

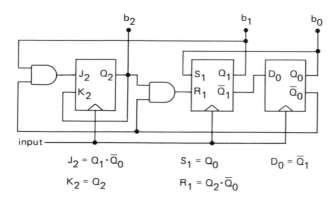

$J_2 = Q_1 \cdot \overline{Q_0}$ $S_1 = Q_0$ $D_0 = \overline{Q_1}$
$K_2 = Q_2$ $R_1 = Q_2 \cdot \overline{Q_0}$

step 2: Write the next-state variables from the flip-flop transition table.

$Q_2^+ = J_2 \cdot \overline{Q_2} + \overline{K_2} \cdot Q_2 = \overline{Q_2} \cdot Q_1 \cdot \overline{Q_0} + \overline{Q_2} \cdot Q_2$
$ = \overline{Q_2} \cdot Q_1 \cdot \overline{Q_0}$
$Q_1^+ = S_1 + \overline{R_1} \cdot Q_1 = Q_0 + (\overline{Q_2} + Q_0) \cdot Q_1$
$ = Q_0 + \overline{Q_2} \cdot Q_1$
$Q_0^+ = D_0 = \overline{Q_1}$

step 3: Complete the state table for the present state and next state determined in step 2.

	present state $Q_2 Q_1 Q_0$	next state $Q_2^+ Q_1^+ Q_0^+$
S_0	0 0 0	0 0 1
S_1	0 0 1	0 1 1
S_2	0 1 1	0 1 0
S_3	0 1 0	1 1 0
S_4	1 1 0	0 0 0
S_5	1 1 1	0 1 0
S_6	1 0 1	0 1 1
S_7	1 0 0	0 0 1

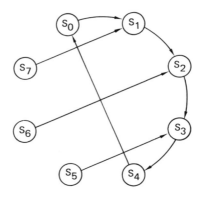

step 4: Draw the state diagram. Hang-up states are two or more unused states that have no next state in the desired sequence. Unless there is an automatic clearing of all flip-flops when the system is energized, it could come up into one of the unused states, never getting into the desired sequence.

step 5: Draw the timing diagram.

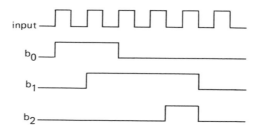

The analysis can be reversed when a circuit is not given.

CONTROL SYSTEMS

FEEDBACK CONTROL LOOPS

Unity Feedback

$C = GE$ = output
R = input
$E = R - C$ = error signal
G = forward transfer function

$$\frac{C}{R} = \frac{G}{1+G}$$

$$\frac{E}{R} = \frac{1}{1+G}$$

$C, E,$ and R are Laplace-transformed variables.

The summing point has negative feedback. Positive feedback is unstable.

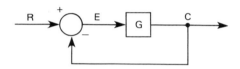

Generalized Control Loops

G_c = compensator transfer function
G_p = plant transfer function
H = feedback loop transfer function

$C = G_c G_p E$
$E = R - HC$

$$\frac{C}{R} = \frac{G_c G_p}{1 + G_c G_p H}$$

$$\frac{E}{R} = \frac{1}{1 + G_c G_p H}$$

$G_c G_p H$ = open loop transfer function of system

All of the above values are complex and expressed in the form $s = \sigma + j\omega$.

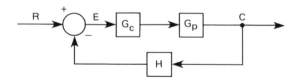

BLOCK DIAGRAM ALGEBRA

Elements in a feedback control loop can be rearranged and combined as shown in the accompanying table.

DC MOTOR POSITIONAL CONTROL SYSTEM EXAMPLE

R_a = armature resistance
R_s = power amplifier output resistance
K_t = tachometer feedback coefficient
L_a = armature inductance
K_v = emf constant
D = friction coefficient in N·m·s
J = mass moment of inertia in N·m·s^2
τ_m = mechanical time constant = $\dfrac{J}{D}$
θ = shaft position in rads
θ_m = maximum shaft position

$$\frac{\theta_c}{V_e} = -A \frac{G_c K}{s\left[1 + \left(\dfrac{2\varsigma}{\omega_n}\right)s + \dfrac{s^2}{\omega_n^2}\right]}$$

$\varsigma > 1 \quad p_1, p_2 = -\varsigma\omega_n \pm \omega_n\sqrt{\varsigma^2 - 1}$
$\varsigma < 1 \quad p_1, p_2 = -\varsigma\omega_n \pm j\omega_n\sqrt{1 - \varsigma^2}$
$\varsigma = 1 \quad p_1, p_2 = -\omega_n$

$$K = \frac{K_v}{K_v^2 + D(R_a + R_s)}$$

$$\omega_n^2 = \frac{K_v^2 + D(R_a + R_s)}{DL_a\tau_m}$$

$$2\varsigma = \frac{\omega_n D[L_a + \tau_m(R_a + R_s)]}{K_v^2 + D(R_a + R_s)}$$

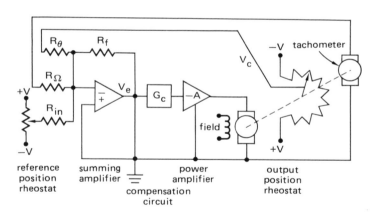

DC motor positional control system with tachometer feedback

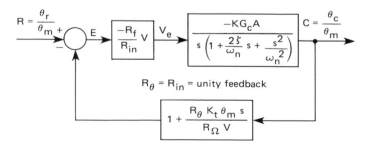

feedback from position rheostat and tachometer

Equivalent Block Diagrams

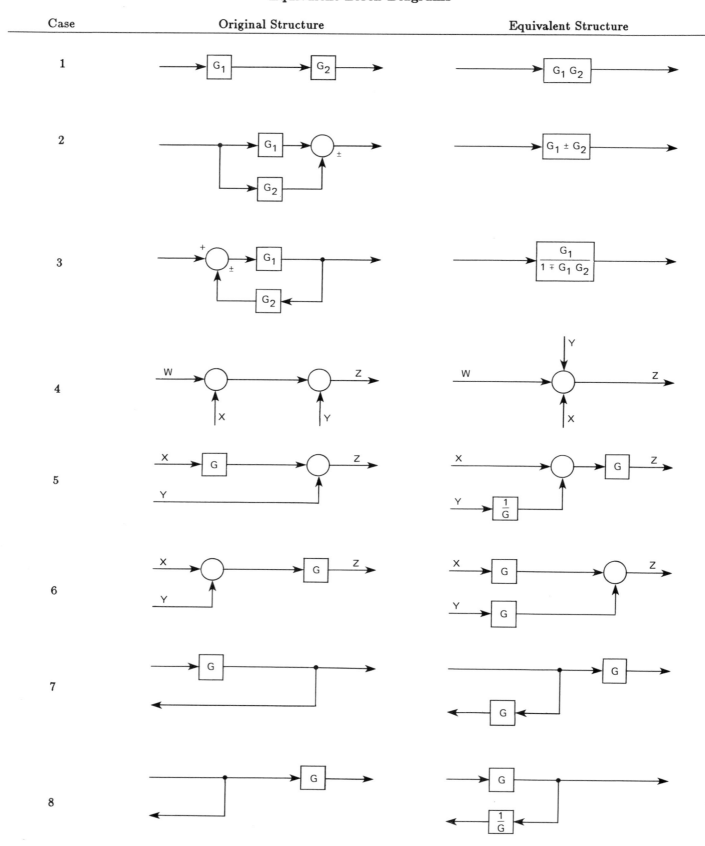

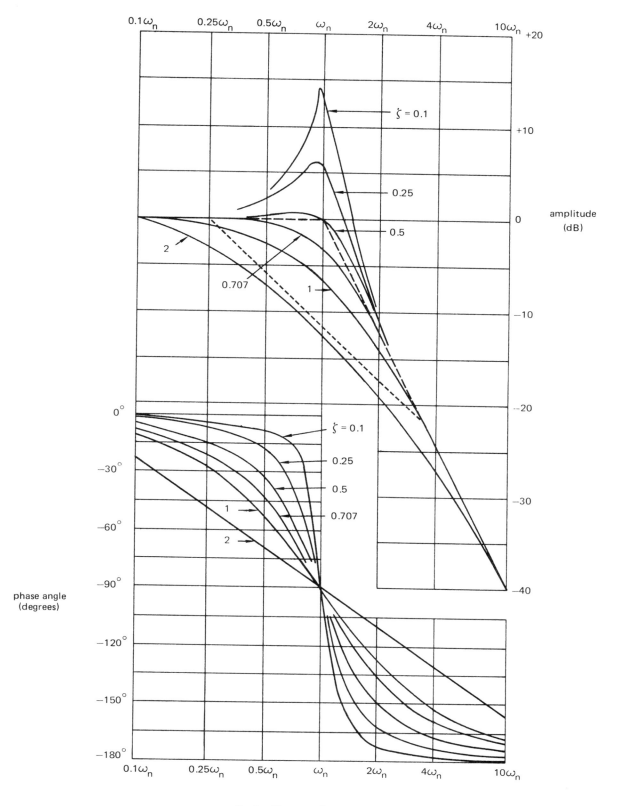

Bode diagram for resonant term

BODE PLOTS

A Bode plot is a plot of decibels versus log to the base 10 of frequency. The asymptotic plots of some transfer functions are shown.

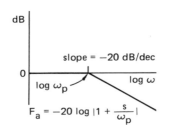
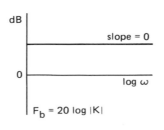
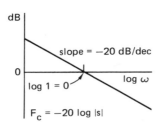
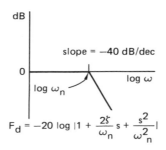

STABILITY

The characteristic equation of a control system is $1 + GH$. For system stability, the function $1 + GH \neq 0$, since $T = G/(1+GH) \to \infty$ as $1+GH \to 0$. Also, $1+GH$ must have no positive real roots since the inverse Laplace transform would result in a positive exponential, $\mathcal{L}^{-1}[1/(s-a)] = e^{at}$. Roots on the $j\omega$ axis are also considered unstable.

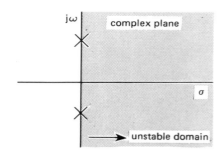

ROUTH-HORWITZ STABILITY

step 1: Reduce the fraction to one with a polynomial numerator and denominator.

$$\frac{C}{R} = \frac{N_g D_h}{D_g D_h + N_g N_h} = \frac{N}{D}$$

step 2: Write D in decreasing powers of s.

$$D = a_n s^n + a_{n-1} s^{n-1} + a_{n-2} s^{n-2} \cdots$$

step 3: If any a_j coefficients have a sign change, the system is unstable.

step 4: Write an array from the decreasing powers of the coefficients.

$$\begin{array}{cccc} s^n & a_n & a_{n-2} & a_{n-4} \cdots \\ s^{n-1} & a_{n-1} & a_{n-3} & a_{n-5} \cdots \\ s^{n-2} & b_{n-1} & b_{n-3} & b_{n-5} \cdots \\ s^{n-3} & b_{n-1} & b_{n-3} & b_{n-5} \cdots \\ \vdots & & & \\ s^0 & & & \end{array}$$

$$b_{n-1} = \frac{\begin{vmatrix} a_n & a_{n-2} \\ a_{n-1} & a_{n-3} \end{vmatrix}}{-a_{n-1}}$$

$$b_{n-3} = \frac{\begin{vmatrix} a_n & a_{n-4} \\ a_{n-1} & a_{n-5} \end{vmatrix}}{-a_{n-1}}$$

step 5: If there are any sign changes in the first column, the system is unstable.

For a given row, if the column 1 term is zero but the column 2 term is finite, replace the column 1 term with the small variable e and continue the test.

If the elements of a row are all zero, replace the zero row elements with elements from a polynomial formed from the elements from the previous row. Continue the test.

ROOT LOCI

A root loci is a plot of points on the s plane where $|GH| = 1$ and angle $\angle GH = \angle 180 \pm n\,360°$.

$$GH = \frac{s-z}{s-p} = |GH| \;\; \angle GH$$

$$\angle GH = \angle s-z - \angle s-p$$

When $GH = -1$, the control system is unstable.

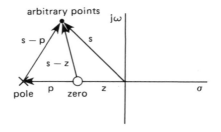

Complex poles and zeros occur in conjugate form and therefore contribute no phase angle to the real axis.

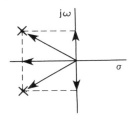

A root locus exists on the real axis when an odd number of poles or zeros exists on the real axis to the right. When two poles or zeros are adjacent, the root loci break away from the real axis.

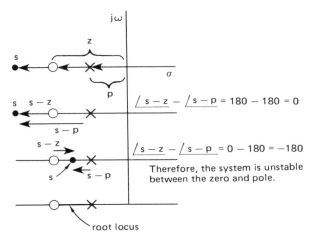

Roots start at poles of GH where K, the gain of GH, $= 0$ and end at the zeros of GH where $K = \infty$.

$$s^2 - ps + K = 0$$

$$s = +\frac{p}{2} \pm \sqrt{\left(\frac{p}{2}\right)^2 - K}$$

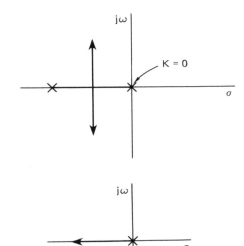

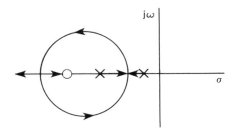

$$\text{radius}^2 = (z - p_1)(z - p_2)$$

STATE VARIABLES

A state variable block diagram is one containing integrators and amplifiers which define the transfer function of a differential equation.

$$\frac{C}{R} = K \frac{(s^n + b_{n-1}s^{n-1} + b_{n-2}s^{n-2} + \cdots)x}{(s^m + a_{m-1}s^{m-1} + a_{m-2}s^{m-2} + \cdots)x}$$

x's are dummy variables.

$$C = K(s^n + b_{n-1}s^{n-1} + b_{n-2}s^{n-2} + \cdots)x$$
$$R = (s^m + a_{m-1}s^{m-1} + a_{m-2}s^{m-2} + \cdots)x$$
$$d^n x = \frac{c(t)}{K} - b_{n-1}d^{n-1}x - b_{n-2}d^{n-2}x + \cdots$$

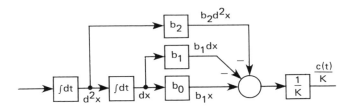

feedforward blocks

Also, $d^m x = R(t) - a_{m-1}d^{m-1}x - a_{m-2}d^{m-2}x + \cdots$

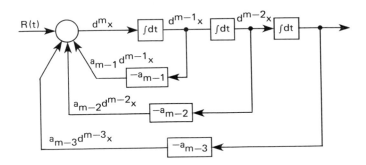

feedback blocks

Feedforward and feedback blocks can be added together to make a complete diagram.

CONTROL SYSTEM DESIGN

To ensure satisfactory transient stability and steady state, a control system phase margin should be no greater than $-150°$ and the gain margin should be 8 to 10 dB.

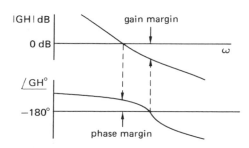

gain and phase margins

Control systems are classified by their ability to follow an input function. The Laplace final value theorem is applied to arrive at an error for the steady state.

$$E_{ss} = \lim_{t=\infty} E(t) = \lim_{s=0} sF(s)$$

n in s^n is the number of integrators. The greater the value of n, the less stable the system.

$$HG(s) = \frac{K\left(1+\frac{s}{z_1}\right)\left(1+\frac{s}{z_2}\right)\cdots}{s^n\left(1+\frac{s}{p_1}\right)\left(1+\frac{s}{p_2}\right)\cdots}$$

System Type	n	Position Error	Velocity Error	Acceleration Error
0	0	$\frac{R}{1+K}$	∞	∞
I	1	0	$\frac{R'}{K}$	∞
II	2	0	0	$\frac{2R''}{K}$

Input Function	Step	Ramp	Parabolic
Laplace of input	$\frac{R}{s}$	$\frac{R'}{s^2}$	$\frac{R''}{s^3}$

ECONOMICS

Cash flow diagrams show receipts and disbursements each year in symbolic form.

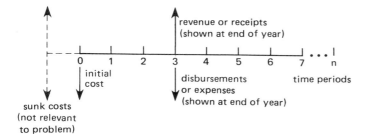

INTEREST RATES

i = effective interest = $\left(1 + \frac{r}{k}\right)^k - 1$

r = nominal rate

k = number of compounding periods per year

The following rules may be used to determine which interest rate is given in a problem:

- Unless specifically qualified in the problem, the interest rate given is an annual rate.
- If the compounding is annually, the rate given is the effective rate. If compounding is other than annually, the rate given is the nominal rate.
- If the type of compounding is not specified, assume annual compounding.

ECONOMIC LIFE

An asset's economic life is the minimum annual cost point of the asset.

PRESENT WORTH METHOD

This method compares two or more alternatives and selects the one with the lowest present cost or greatest present revenue.

CAPITAL COST METHOD

capital cost = initial cost + $\dfrac{\text{annual costs}}{i}$

ANNUAL COST METHOD

This method selects the alternative with the lowest equivalent annual cost.

BENEFIT COST RATIO METHOD

$$\frac{\text{present worth of all benefits}}{\text{present worth of all costs}} \geq 1$$

COST ACCOUNTING

Here are some important points concerning basic cost accounting:

The sum of direct material and direct labor costs is known as the *prime cost*.

Factor Name	Converts	Symbols	Formula
single payment compound amount	P to F	$(F/P, i\%, n)$	$(1+i)^n$
present worth	F to P	$(P/F, i\%, n)$	$(1+i)^{-n}$
uniform series sinking fund	F to A	$(A/F, i\%, n)$	$\dfrac{i}{(1+i)^n - 1}$
capital recovery	P to A	$(A/P, i\%, n)$	$\dfrac{i(1+i)^n}{(1+i)^n - 1}$
compound amount	A to F	$(F/A, i\%, n)$	$\dfrac{(1+i)^n - 1}{i}$
equal series present worth	A to P	$(P/A, i\%, n)$	$\dfrac{(1+i)^n - 1}{i(1+1)^n}$
uniform gradient	G to P	$(P/G, i\%, n)$	$\dfrac{(1+i)^n - 1}{i^2(1+i)^n} - \dfrac{n}{i(1+i)^n}$

Indirect costs may be called *indirect manufacturing expenses* (IME).

Indirect costs may also include the overhead sector of the company (e.g., secretaries, engineers, and corporate administration). In this case, the indirect cost is usually called *burden* or *overhead*. Burden may also include the *equivalent uniform annual cost* (EUAC) of non-regular costs which must be spread evenly over several years.

The cost of a product is usually known in advance from previous manufacturing runs or by estimation. Any deviation from this known cost is called a *variance*. Variance may be broken down into *labor variance* and *material variance*.

Indirect cost per item is not easily measured. The method of allocating indirect costs to a product is as follows:

step 1: Estimate the total expected indirect (and overhead) costs for the upcoming year.

step 2: Decide on some convenient vehicle for allocating the overhead to production. Usually, this vehicle is either the number of units expected to be produced or the number of direct hours expected to be worked in the upcoming year.

step 3: Estimate the quantity or size of the overhead vehicle.

step 4: Divide expected overhead costs by the expected overhead vehicle to obtain the unit overhead.

step 5: Regardless of the true size of the overhead vehicle during the upcoming year, one unit of overhead cost is allocated per product.

Quick – I need additional study materials!

Please rush me the review materials I have checked. I understand any item may be returned for a full refund within 30 days. I have provided my bank card number as method of payment, and I authorize you to charge your current prices against my account.

For the E-I-T Exam: Solutions Manuals:
[ ] Engineer-In-Training Review Manual   [ ]
    [ ] Engineering Fundamentals Quick Reference Cards
    [ ] E-I-T Mini-Exams

For the P.E. Exams:
[ ] Civil Engineering Reference Manual   [ ]
    [ ] Civil Engineering Sample Examination
    [ ] Civil Engineering Quick Reference Cards
    [ ] Seismic Design
    [ ] Timber Design
    [ ] Structural Engineering Practice Problem Manual
[ ] Mechanical Engineering Reference Manual   [ ]
    [ ] Mechanical Engineering Quick Reference Cards
    [ ] Mechanical Engineering Sample Examination
[ ] Electrical Engineering Reference Manual   [ ]
[ ] Chemical Engineering Reference Manual   [ ]
    [ ] Chemical Engineering Practice Exam Set
[ ] Land Surveyor Reference Manual   [ ]

Recommended for all Exams:
[ ] Expanded Interest Tables
[ ] Engineering Law, Design Liability, and Professional Ethics

SHIP TO:

Name _____

Company _____

Street _____ Apt. No. _____

City _____ State _____ Zip _____

Daytime phone number _____

CHARGE TO (required for immediate processing):

_____ _____
    VISA/MC/AMEX account number   expiration date

_____
    name on card

_____
    signature

Send more information

Please send me descriptions and prices of all available E-I-T and P.E. review books. I understand there will be no obligation.

_____
_____
_____
_____

A friend of mine is taking the exam too. Send additional literature to:

_____
_____
_____
_____

I disagree...

I think there is an error on page _____. Here is the way I think it should be.

Title of this book: **ELECTRICAL ENGINEERING QUICK REFERENCE CARDS**

[ ] Please tell me if I am correct.

Contributed by (optional):

_____
_____
_____

BUSINESS REPLY MAIL

FIRST CLASS PERMIT NO. 33, BELMONT, CA

POSTAGE WILL BE PAID BY ADDRESSEE

PROFESSIONAL PUBLICATIONS, INC.
1250 Fifth Avenue
Belmont, CA 94002-9900

NO POSTAGE NECESSARY IF MAILED IN THE UNITED STATES

BUSINESS REPLY MAIL

FIRST CLASS PERMIT NO. 33, BELMONT, CA

POSTAGE WILL BE PAID BY ADDRESSEE

PROFESSIONAL PUBLICATIONS, INC.
1250 Fifth Avenue
Belmont, CA 94002-9900

NO POSTAGE NECESSARY IF MAILED IN THE UNITED STATES

BUSINESS REPLY MAIL

FIRST CLASS PERMIT NO. 33, BELMONT, CA

POSTAGE WILL BE PAID BY ADDRESSEE

PROFESSIONAL PUBLICATIONS, INC.
1250 Fifth Avenue
Belmont, CA 94002-9900

NO POSTAGE NECESSARY IF MAILED IN THE UNITED STATES